United States
Department of
Agriculture

Forest Service

Pacific Northwest
Forestand Range
Experiment Station

General Technical
Report
PNW -112

August 1980

Wetwood in Trees:
A TimberResource
Problem

J.C. Ward and W .Y. Pong

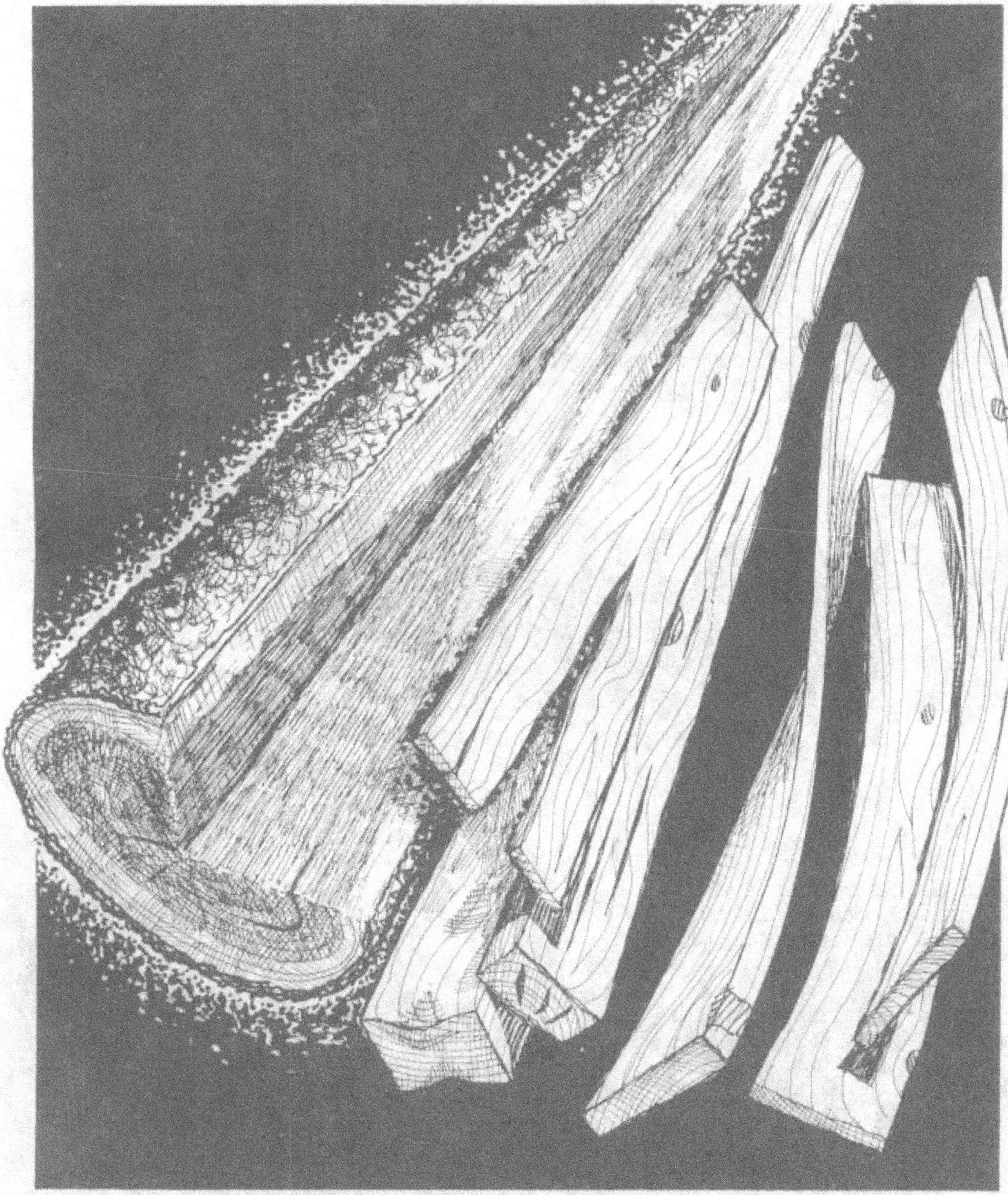

Abstract

Contents

Available information on wetwood is presented. Wetwood is a type of heartwood which has been internally infused with water. Wetwood is responsible for substantial losses of wood, energy and production expenditures in the forest products industry.

The need to evaluate these losses and to find ways to eliminate or minimize them is emphasized.

Because of increased interest in wetwood, excellent opportunities exist in initiating a comprehensive program of research. Both short and long-term studies are included in the program with the short-term studies answering the immediate utilization problems created by wetwood and the long-term studies directed at examining the causes of wetwood and its control and prevention.

Metric Equivalents

1 inch = 2.54 centimeters
1 foot = 0.304 8 meter
1 cubic foot = 0.028 32 cubic meter
1 pound = 0.453 6 kilogram
1 pound/square inch = 0.070 38 kilogram/ square centimeter
$5/9(°F-32) = °C$

AUTHORS

J.C. Ward is research forest products technologist, U.S. Forest Products Laboratory, Madison, Wisconsin.

W.Y. Pong is research forest products technologist, Pacific Northwest Forest and Range Experiment Station, Portland, Oregon.

Introduction

This paper summarizes available information on the timber quality characteristic described as wetwood. Included are some possible causes of wetwood, where it occurs in trees, its properties, and the problems associated with it. The need for research to solve utilization problems associated with wetwood and the processing of logs and products containing it is recognized.

With this report as a guide, a concerted research effort directed at the problem of wetwood can be organized. Such research will, in the short run, target in on and provide answers to wetwood-related processing problems presently plaguing industry and, at the same time, build a base of scientific knowledge with which to solve the problem of wetwood in timber stands.

Characteristics of Wetwood

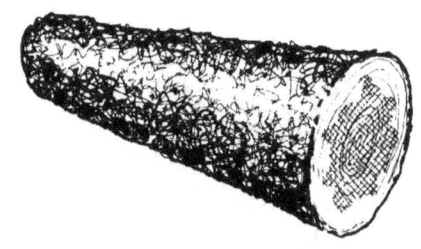

Wetwood is a type of heartwood in standing trees which has been internally infused with water (64, 79, 142). Wood infused with water from an external source (e.g., logs exposed to spray, rain, or storage ponds) does not fit this definition. In some tree species, wetwood often has a water-soaked translucent appearance and is variously designated as water core, sinker heart, wet core, wet heart, and discolored wood. When frozen, the water-soaked wetwood appears as a distinct, hard, glossy surface on the ends of winter-cut logs of Scots pine (Pinus sylvestris L.) (117) and balsam fir (164).[1]

In other species, the typical water-soaked appearance may be absent. In this case, wetwood has the appearance of either normal heartwood or it has an unusually dark color. This dark color leads to the term "false" or "pathological" heartwood. Red heart in white or paper birch is a type of wetwood[2] (31, 68). One type of wetwood in white fir is called blackheart (146).

---

[1]/Scientific names of commercial North American tree species mentioned in this report are listed in table 1 (conifers), page 5, and table 2 (hardwoods), page 6.

[2]/R. W. Davidson, Colorado State University, personal communication to J. C. Ward.

Causes of Wetwood Formation

In general, wetwood is higher in moisture content than the adjacent normal heartwood. Reports of unusually high moisture content in heartwood may result from an investigator's inability to recognize wetwood. In comparison with sapwood, wetwood can be higher in moisture content ([67], [79], [106], [117], [197]) or lower or equal in moisture ([20], [48], [121], [133], [203], [212], [215]). Regardless of external appearance, wetwood differs from normal wood in physical and chemical properties and generally is more difficult to dry.

Wetwood in trees has been attributed to a number of causes: microbial (bacterial), nonmicrobial (injury), and normal age-growth formation. Associative relationships have been the basis of many evaluations of the causes of wetwood; actual tests to confirm cause-and-effect relationships have had little success.

Much of the research on wetwood has been directed toward investigating the association of bacterial infection and the occurrence of wetwood in trees ([20], [25], [46], [49], [60], [78], [79], [93], [106], [158], [167], [168], [190], [191], [202], [203], [206], [207], [214], [217], [219]). Not all of these studies have found bacteria to be associated with wetwood. Bacteria have frequently been isolated from healthy sapwood and heartwood of living trees ([12], [13], [40], [107], [144], [190], [214]).

The concept that wetwood is the result of a mixed population or invasion of more than one species of bacteria could explain some of the differences in results. This concept would be analogous to the successions of organisms in the development of wood decay and associated discolorations described by Shigo ([171]) and Shigo and Hillis ([173]). Strictly anaerobic bacteria as well as facultative anaerobes have been isolated from wetwood ([158], [174], [199], [202], [203], [205], [206], [222]). Laboratory techniques for isolating andculturing strictly anaerobic bacteria are different from those for aerobic and facultative anaerobes and fungi. This could explain why investigative results do not always agree about the relative importance of bacteria to formation of wetwood.

That wetwood formation or initiation may be nonmicrobial in nature has been suggested by some investigators ([20], [45], [107], [144]) and implied by others ([12], [40], [190]). These investigations conclude that bacteria are part of an indigenous microflora of normal wood, living in a viable but quiescent state until some change occurs in the wood to provide a favorable substrate (wetwood) for the bacterial growth. Bacteria isolated from wetwood of poplar utilized capillary liquid from sapwood and heartwood for growth ([190]).

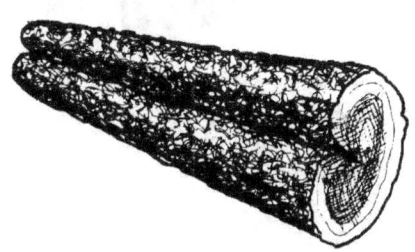

Wetwood has been associated with physical, mechanical, and biological injuries[3/] ([20], [79], [186]). Whether these injuries result in wetwood formation or initiate it has not been substantiated. Field observations have shown that the normal cylindrical form of wetwood in white fir deviated radially and longitudinally to regions of natural and silvicultural injuries (observations by W. Y. Pong). Similar deviations have been noted in conjunction with injuries apparently caused by insect attacks[4/] ([78], [79], [211]) and stem cankers of dwarf mistle-

toe in true firs[5/] ([216]). A fairly consistent association between wetwood and decay has been noted in eastern Oregon true firs/ and other tree species ([79]). We have observed that internal heart rot in tree stems often has a peripheral ring of sound wetwood, whereas stems with wetwood do not necessarily have associated decay. There is some belief that wetwood may inhibit decay ([93]).

The inability to consistently associate wetwood with micro-organisms has led to the conclusion that wetwood in conifers is only a condition of excessive accumulation of moisture and not a symptom of disease ([142], [165]). There are opposing views on the source of the excessive moisture in wetwood. Some suggest direct entry of atmospheric moisture through stem openings, such as dead branch stubs ([20]); others believe water in wetwood has an internal origin from moisture sources in the root and stem ([45], [78]). In Japan the occurrence of wet heartwood is considered normal for some species of hardwoods and abnormal for conifers ([220]). Whether formation of wetwood is the result of a natural process of the tree or is bacterial in origin is discussed at length by Hartley, et al. ([79]). Until additional evidence is available, we will consider wetwood the manifestation of a syndrome of abnormal physiological events in the living tree and not a specific disease.

[3/]W. Y. Pong. 1967. Preliminary study of grade defects in true firs. Interim rep. USDA For. Serv. Pac. Southwest For. and Range Exp. Stn., 31 p. Berkeley, Calif. (Unpublished report on file at Pac. Northwest For. and Range Exp. Stn., Portland, Oreg.)

[4/] Personal communication to W. Y. Pong from K. R. Shea, Assistant Director, Science and Education Administration, U.S. Department of Agriculture, Washington D.C., and from W. W. Wilcox, Forest Products Laboratory, University of California, Richmond.

[5/] J. R. Parmeter, Department of Plant Pathology, University of California, Berkeley, personal communication to W. Y. Pong.

[6/]Field observations by P. E. Aho, Forestry Sciences Laboratory, Corvallis, Oreg.

Wetwood in Trees and Logs

Occurrence

Wetwood occurs in both conifers and hardwoods, but its frequency can vary by species, age, and growing conditions of trees. Occurrence of wetwood in the more important commercial timber species of North America is presented in table 1 for softwoods and table 2 for hardwoods. Each species is listed by its apparent susceptibility to wetwood formation. These classifications are derived from our observations, personal communications from colleagues, office reports, and the published literature. Tree species in which wetwood either has never been reported or is rarely observed are placed in class 1. Class 3 includes the most susceptible tree species or those that can be expected to develop wetwood either with age or under less than optimum growing conditions. The intermediate class 2 contains species in which wetwood is found in one locality and rarely, if at all, in another region.

For some species in both tables 1 and 2, the information is tentative; as additional field information becomes available a species may be moved from one class to another. Sufficient information exists to indicate that wetwood will probably develop in hemlocks, true firs, and cottonwoods from most regions. For species such as white pine, red oak, and maple, wetwood is found in many trees of one region and completely lacking in trees of another region. Thus, if the rating of a species in tables 1 and 2 is based on reports from a limited area of the total range of the species, then its status could be reversed when additional information is available.

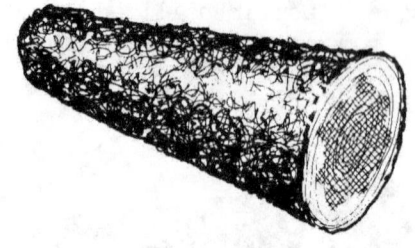

Not all the tree species in class 3 are affected with wetwood to the same degree or frequency. There are differences within a genus; for example, white fir and grand fir are affected more than are noble fir and Pacific silver fir. From his studies with white fir in northern California, Wilcox[7] found that all sample trees, including young ones, contained some wetwood. This appears to be the rule for white fir in northern California, but in some stands wetwood appears to be rare or nonexistent.[8] The presence of wetwood within the hardwood genus *Populus* appears to be the rule, but again there can be exceptions. Some investigators could not find eastern cottonwood trees that were free or wetwood[9] (191, 222). For European white poplar (*Populus alba* L.), the presence of wetwood is the rule in pistillate trees and the exception in staminate trees[10] (71).

7/ W. W. Wilcox, Forest Products Laboratory, University of California, Richmond, personal communication to W. Y. Pong.

8/J. C. Ward, unpublished data on file at U.S. Forest Products Laboratory, Madison, Wis.

9/J. B. Baker, Southern Forest Experiment Station, New Orleans, La., personal communication to J. C. Ward.

10/J. C. Ward and J. G. Zeikus, unpublished data on file at U.S. Forest Products Laboratory, Madison, Wis.

Table 1--Occurrence of wetwood in commercially important conifers of the United States[1]

Species	Frequency class[2]		
	1 Occasional or none	2 Scattered prevalence	3 Generally prevalent
Douglas-fir (Pseudotsuga menziesii (Mirb.) Franco)	+		
Eastern hemlock (Tsuga canadensis (L.) Carr.)			+
Western hemlock (Tsuga heterophylla (Raf.) Sarg.)			+
Mountain hemlock (Tsuga mertensiana (Bong.) Carr.)			+
True fir:			
Balsam fir (Abies balsamea (L.) Mill.)			+
White fir (Abies concolor (Gord. & Glend.) Lindl. ex Hildebr.)			+
Grand fir (Abies grandis (Dougl.) ex D. Don) Lindl.)			+
Subalpine fir (Abies lasiocarpa (Hook.) Nutt.)			+
Pacific silver fir (Abies amabilis (Dougl.) ex Forbes)		+	
Noble fir (Abies procera Rehd.)		+	
California red fir (Abies magnifica A. Murr.)		+	
Shasta red fir (Abies magnifica var. shastensis Lemm.)		+	
Spruce :			
White spruce (Picea glauca (Moench) Voss)	+		
Black spruce (Picea mariana (Mill.) B.S.P.)	+		
Red spruce (Picea rubens Sarg.)	+		
Engelmann spruce (Picea engelmannii Parry ex Engelm.)	+		
Sitka spruce (Picea sitchensis (Bong.) Carr.)	+		
Soft pine:			
Eastern white pine (Pinus strobus L.)		+	
Western white pine (Pinus monticola Dougl. ex D. Don)		+	
Sugar pine (Pinus lambertiana Dougl.)		+	
Hard pine:			
Jack pine (Pinus banksiana Lamb.)	+		
Lodgepole pine (Pinus contorta Dougl. ex Loud.)		+	
Red pine (Pinus resinosa Ait.)	+		
Ponderosa pine (Pinus ponderosa Dougl. ex Laws.)		+	
Shortleaf pine (Pinus echinata Mill.)	+		
Loblolly pine (Pinus taeda L.)	+		
Longleaf pine (Pinus palustris Mill.)	+		
Slash pine (Pinus elliottii Engelm. var. elliottii)	+		
Jeffrey pine (Pinus jeffreyi Grev. & Balf.)	+		
Western larch (Larix occidentalis Nutt.)		+	
Tamarack (Larix laricina (Du Roi) K. Koch)		+	
Redwood (Sequoia sempervirens (D. Don) Endl.)		+	
Giant sequoia (Sequoia dendron giganteum (Lindl.) Buchholz)	+		
Baldcypress (Taxodium distichum (L.) Rich.)	+		
Cedar:			
Incense-cedar (Libocedrus decurrens Torr.)	+		
Western redcedar (Thuja plicata Donn ex D. Don)		+	
Northern white-cedar (Thuja occidentalis L.)	+		
Alaska-cedar (Chamaecyparis nootkatensis (D. Don) Spach)	+		
Port-Or ford-cedar (Chamaecyparis lawsoniana (A. Murr.) Parl.)	+		
Atlantic white-cedar (Chamaecyparis thyoides (L.) B.S.P.)	+		
Eastern redcedar (Juniperus virginiana L.)	+		
Juniper:			
Alligator juniper (Juniperus deppeana Steud.)	+		
Western juniper (Juniperus occidentalis Hook.)	+		

[1]/From field observations by authors augmented with personal communications; also from literature, particularly Hartley et al. (79) and Lutz (126).

[2]/Frequency class 1 includes tree species in which wetwood rarely, if ever, occurs and species in which wetwood has not been observed, reported, or recognized as wetwood. Class 2 includes species in which wetwood will develop on some sites and not on other sites, together with species that tend to develop wetwood with increasing age on certain sites. Class 3 includes species very susceptible to formation of wetwood on most sites and in young- as well as old-growth timber.

Table 2--Occurrence of wetwood in commercially important hardwoods of the United States[1]

Species	Frequency class[2]		
	1 Occasional or none	2 Scattered prevalence	3 Generally prevalent
Red oak group:			
Northern red oak (Quercus rubra L.)		+	
Black oak (Quercus velutina Lam.)		+	
Scarlet oak (Quercus coccinea Muenchh.)		+	
Pin oak (Quercus palustris Muenchh.)		+	
Water oak (Quercus nigra L.)		+	
Nuttall oak (Quercus nuttalli Palmer)	+		
Southern red oak (Quercus falcata Michx. var. falcata)	+		
Cherrybark oak (Quercus falcata var. pagodaefolia Ell.)		+	
Willow oak (Quercus phellos L.)	+		
Laurel oak (Quercus laurifolia Michx.)		+	
Shumard oak (Quercus shumardii Buckl. var. shumardii)	+		
California black oak (Quercus kelloggii Newb.)		+	
White oak group:			
White oak (Quercus alba L.)	+		
Bur oak (Quercus macrocarpa Michx.)	+		
Overcup oak (Quercus lyrata Walt.)		+	
Swamp white oak (Quercus bicolor Willd.)		+	
Chestnut oak (Quercus prinus L.)	+		
Swamp chestnut oak (Quercus michauxii Nutt.)	+		
Chinkapin oak (Quercus muehlenbergii Engelm.)	+		
Post oak (Quercus stellata Wangenh. var. stellata)		+	
Oregon white oak (Quercus garryana Dougl. ex Hook)		+	
California white oak (Quercus lobata Nee)		+	
Tanoak (Lithocarpus densiflorus (Hook. & Arn.) Rehd.)		+	
Hickory :			
Bitternut hickory (Carya cordiformis (Wangenh.) K. Koch)		+	
Water hickory (Carya aquatica (Michx. f.) Nutt.)		+	
Pecan (Carya illinoensis (Wangenh.) K. Koch)		+	
Shagbark hickory (Carya ovata (Mill.) K. Koch)	+		
Shellbark hickory (Carya laciniosa (Michx. f.) Loud.)	+		
Mockernut hickory (Carya tomentosa (Poir.) Nutt.)	+		
Pignut hickory (Carya glabra (Mill.) Sweet)	+		
Black walnut (Juglans nigra L.)		+	
Butternut (Juglans cinerea L.)	+		
Sweetgum (Liquidambar styraciflua L.)		+	
Maple:			
Sugar maple (Acer saccharum Marsh.)	+		
Silver maple (Acer saccharinum L.)	+		
Red maple (Acer rubrum L.)		+	
Black maple (Acer nigrum Michx. f.)	+		
Bigleaf maple (Acer macrophyllum Pursh)	+		
Black tupelo (Nyssa sylvatica Marsh.)		+	
Water tupelo (Nyssa aquatica L.)		+	
Yellow poplar (Liriodendron tulipifera L.)		+	
Magnolias (Magnolia L. spp.)	+		

Table 2--Continued

Species	Frequency Class[2]		
	1 Occasionial or none	2 Scattered prevalence	3 Generally prevalent
Ash:			
White ash (Fraxinus americana L.)		+	
Green ash (Fraxinus pennsylvanica Marsh.)		+	
Black ash (Fraxinus nigra Marsh.)		+	
Oregon ash (Fraxinus latifolia Benth.)		+	
American beech (Fagus grandifolia Ehrh.)		+	
Poplar and aspen:			
Eastern cottonwood (Populus deltoides Bartr. ex Marsh.)			+
Plains cottonwood (Populus sargentii Dode)			+
Narrowleaf cottonwood (Populus angustifolia James)			+
Black cottonwood (Populus trichocarpa Torr. & Gray)			+
Fremont cottonwood (Populus fremontii Wats.)			+
Swamp cottonwood (Populus heterophylla L.)			+
Balsam poplar (Populus balsamifera L.)			+
Quaking aspen (Populus tremuloides Michx.)		+	
Largetooth aspen (Populus grandidentata Michx.)		+	
Black willow (Salix nigra Marsh.)			+
Red alder (Alnus rubra Bong.)		+	
Birch :			
Paper birch (Betula papyrifera Marsh.)		+	
Gray birch (Betula populifolia Marsh.)		+	
Yellow birch (Betula alleghaniensis Britton)		+	
River birch (Betula nigra L.)		+	
American sycamore (Platanus occidentalis L.)			+
Black cherry (Prunus serotina Ehrh.)		+	
Northern catalpa (Catalpa speciosa Warder ex Engelm.)	+		
Elm:			
American elm (Ulmus americana L.)		+	
Slippery elm (Ulmus rubra Mühl.)		+	
Rock elm (Ulmus thomasii Sarg.)	+		
Hackberry (Celtis occidentalis L.)	+		
Sugarberry (Celtis laevigata Willd.)	+		
Ohio buckeye (Aesculus glabra Willd.)	+		
Yellow buckeye (Aesculus octandra Marsh.)	+		
American basswood (Tilia americana L.)		+	
White basswood (Tilia heterophylla Vent.)	+		
Pacific madrone (Arbutus menziesii Pursh)		+	
Red mulberry (Morus rubra L.)		+	
Black locust (Robinia pseudoacacia L.)	+		
Honey locust (Gleditsia triacanthos L.)		+	
Water locust (Gleditsia aquatica Marsh.)		+	
Kentucky coffeetree (Gymnocladus dioicus (L.) K. Koch)		+	

[1]From field observations by authors augmented with personal communications; also from literature, particularly Hartley et al. (79) and Lutz (126).

[2]Frequency class 1 includes tree species in which wetwood rarely, if ever, occurs and species in which wetwood has not been observed, reported, or recognized as wetwood. Class 2 includes species in which wetwood will develop on some sites and not on other sites, together with species that tend to develop wetwood with increasing age on certain sites. Class 3 includes species very susceptible to formation of wetwood on most sites and in young- as well as old-growth timber.

Even though considerable data may show that a species in class 1 is not susceptible to wetwood formation, it may be important to continue looking for exceptions. For example, we have rarely found wetwood in Douglas-fir and then it was limited in volume to small streaks or pockets. Nevertheless, one small stand of Douglas-fir timber in the Cascade Range of Oregon contained extensive wetwood in the logs, and the lumber was difficult to dry to a uniform moisture content.[11] In Washington, a Douglas-fir bridge timber which failed from ring shake was found to have contained extensive bacterial wetwood (see footnote 8).

Age has some influence on formation of wetwood. In general, there is less wetwood in young trees: some foresters think that the problem will be solved when the last old-growth timber is cut. In some species, such as cottonwood, aspen, elm, maple, and white fir, wetwood has been recorded in very young trees (see footnote 7) (79).

When wetwood is prevalent in young trees, site or cultural practices may be the major contributing factors. Field observations by the senior author indicate that site conditions may influence the incidence of wetwood in young-growth western hemlock. Wallin (197) observed a relation between wetwood formation in balsam poplar and soil type. Wetwood is most prevalent in the lower stems of western redcedar growing on wet, swampy sites (73, 89). These soil or site factors may favor increases in populations of micro-organisms associated with wetwood. Similarly, cultural practices could contribute to the establishment and growth of these micro-organisms in trees. Most cottonwood trees grown in plantations on good soils in the Mississippi delta develop wetwood within 2 years. This is apparently related to the

planting of scions in moist soil (see footnote 9). Second-growth aspen in northern Minnesota contains more wetwood than aspen from the original forest.[12] This may be due to the origin of the second-growth trees from sprouts and a different species composition of the second-growth stands favoring attacks by both micro-organisms and stem-boring insects. Eis (53) found that root grafts in natural stands of Douglas-fir, western hemlock, and western redcedar can be a major factor in the spread of decay organisms from stumps to neighboring trees. Root grafts may also be a factor in the initiation and spread of wetwood. Cole and Streams (42) suggest that wetwood bacteria are spread by slime-flux insects.

Types of Wetwood and Patterns Within the Tree

With few exceptions, formation of wetwood has been restricted to the heartwood and heartwood-sapwood transition zones of living trees. Essentially two types of wetwood can occur in the living tree: wetwood formed from the aging of normal sapwood nearest the heartwood or injured sapwood and wetwood developed in previously formed heartwood. The two types frequently occur together and are difficult to distinguish.

11/C. J. Kozlik, Department of Forest Products, Oregon State University, Corvallis, personal communication to J. C. Ward.

12/E. M. Ballman, Chief Forester, Diamond International Corp., Cloquet, Minn., personal communication to J. C. Ward.

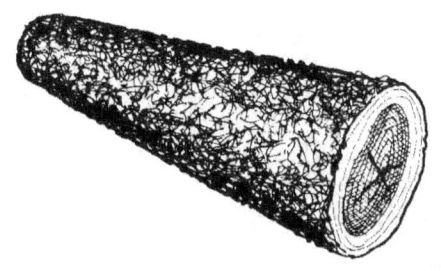

Although the spatial distribution of wetwood in standing trees will vary by point of origin, duration of formation, and other such factors, there are essentially three general patterns. Most common is the conical pattern in the central core of the lower bole--the wetwood originates at injuries in the roots OK at the root collar and tapers to an apex in the upper bole. This pattern has been found in true firs (45, 92, 147, 215), western hemlock (203), northern red oak (75, 202), eastern cottonwood (see footnote 9), and western larch (see footnote 8).

Wetwood is generally not found in the outer sapwood of living trees; reports to the contrary indicate that such occurrence is associated with insect attack, stem cankers, and wounds of various forms (79, 91; also see footnotes 3, 5, and 7). In these instances, the wetwood zone of the central core deviated radially and longitudinally to the traumatized zone of the sapwood. Knutson (106), however, was not always able to relate wetwood in the sapwood of aspen to visible wounds, but he considered wetwood to be generally a phenomenon of the sapwood. Takizawa et al. (186) found from the cytology of wetwood in the Japanese fir Todomatsu (Abies sachalinensis Mast.) that wetwood was more comparable to sapwood than heartwood but it existed as streaks in the outer heartwood.

It is important that a distinction is made between the wetwood OK "sinker" stock (i.e., wood that will not float) formed in the standing tree and the "sinker wetwood" resulting from storage of logs in water. In pond-stored logs, the affected wood is mainly the sapwood and the wood of the central core exposed on the log ends. Pond-stored logs cease to float because of increased permeability from bacterial disintegration of pit membranes in the sapwood but not the heartwood (44, 50, 55, 69, 72, 90, 95, 105, 119, 157, 180, 195, 208). In comparison, wetwood in standing trees is confined more to the heartwood and has registered lower permeability than the adjacent normal wood, particularly sapwood. The permeability of wetwood from trees is discussed in the section on properties of wetwood.

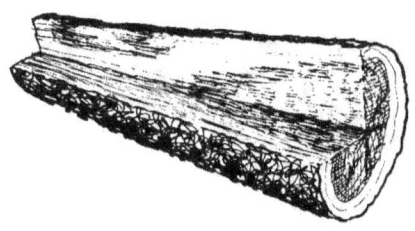

The second pattern consists of streaks OK columns of wetwood which can be traced to injuries or to stubs of dead branches in the upper stem. The basal portion of these trees may be free of wetwood. Upper stem wetwood has been noted in trembling aspen (106), cricket-bat willow (Salix alba var. calva G. F. W. Mey.) (49), elm (33), eastern white pine (123), silver fir (Abies alba Mill) (20), California white fir,[13] and eastern and western hemlock.[14] (see footnote 11).

[13]/A. L. Shigo, Forestry Sciences Laboratory, Durham, N. H., personal communication to J. C. Ward.

[14]/ Field observations by the authors.

Wetwood Properties--
A Microbial Implication

The third pattern appears when a tree contains both basal and upper stem wetwood formations that extend and coalesce into a single massive formation. This pattern was noted in Scots pine and Norway spruce (Picea abies (L.) Karst) by Lagerberg (117) who also was the first to describe the other patterns. It is not uncommon for aspen in Minnesota and Wisconsin (see footnote 8) and balsam fir (59) to have the combined wetwood pattern.

Wetwood has distinctive physical and chemical properties. These properties appear to result from microbial action on woody tissue in the living tree, but more comprehensive tests are needed to validate this relationship.

Odor

The odor of wetwood in the green condition differs from that of normal wood and is reminiscent of fermentation processes. This indicates an influence by anaerobic bacteria. Odors in wetwood ranging from rancid to a fetid rumenlike odor have been traced to bacterial metabolism of woody components, particularly extractives and hemicelluloses (1, 187, 222, 223). Using potato mash media in the laboratory, the senior author (unpublished data) was able to reproduce the rancid, fatty acid odors of wetwood with pure cultures of Clostridium spp. isolated from wetwood of cottonwood, red oak, and white fir. Bauch et al. (20) found acetic acid, propionic acid, and n-butyric acid present in the wetwood of silver fir, but not in the sapwood. They believe these fatty acids indicate a high bacterial activity in wetwood.

Hydrogen Ion Concentration or pH

The prevailing belief is that wetwood has a higher pH than normal wood. This can probably be traced to the publication, "Wetwood, Bacteria, and Increased pH in Trees," by Hartley et al. (79), who cite results indicating a higher pH in wetwood of many hardwoods and conifers. They cautioned, however, that not all wetwood is notably high in pH and that it is unsafe to assume that all wood with high pH is wetwood or has been.

A review of published data indicates that a comprehensive statement concerning the pH of wetwood cannot be made now. Reported pH values are generally on the alkaline side for wetwood in hardwoods and on the acid side for conifers, but there are exceptions between tree species and even within single species and trees. In comparison with normal sapwood and heartwood, wetwood of European aspen (Populus tremula L.) was more acid (144), but wetwood in aspen from Minnesota can be either more acid or more alkaline (41, 67, 106). All wetwood in elm tested by Carter (33) was more alkaline than a slightly acidic normal wood. Seliskar (167) found that wetwood in elm can be either acid or alkaline, whereas the wetwood of Lombardy poplar (Populus nigra var. italica Muenchh.) was consistently on the alkaline side. Wetwood in balsam poplar is slightly basic, the sapwood slightly acidic, and the heartwood essentially neutral (197). Red heart, a type of wetwood in paper birch, is alkaline adjacent to the acidic sapwood and turns acidic near the pith (31). Wetwood in conifers is more acid than the adjacent normal wood for California white fir (212, 215), silver fir (20), and western hemlock (111). Another study of western hemlock reported that wetwood and normal heartwood had a similar range in pH values from 4.1 to 5.5 (165). Any increases in the pH of wetwood from conifers are small (79).

Until additional studies are made we can only speculate on the causes for differences in pH between wetwood and normal wood and between wetwood in individual trees and species. Microbial populations in trees are possible causative agents, but they must be considered in conjunction with site and chemistry of the trees. Seliskar (167) noted that elms on poor sites had alkaline wetwood, whereas wetwood from trees on good sites was acidic. From laboratory test cultures, Carter (33) observed that the elm wetwood bacterium, Erwinia nimipressuralis, will turn a nutrient broth containing a sugar strongly acid, but growth on nutrient broth alone will result in an alkaline pH. Toole (191) reports that wetwood in eastern cottonwood growing along the Mississippi River between Vicksburg, Mississippi, and Memphis, Tennessee, is neutral to alkaline, and the sapwood is slightly acidic. The senior author found similar acidic pH values for sapwood of cottonwood growing along the Mississippi River in Wisconsin, but the wetwood had either higher or lower pH values than sapwood. The composition of the microbial population in wetwood varies with either acid or alkaline pH conditions for eastern cottonwood (222) and California white fir (see footnote 10). Whether differences in pH are the cause or the result of differences in microbial populations has not been determined.

Anaerobiosis

Results from analysis of gas samples from wetwood in hardwood trees range from near anaerobic conditions in elm (33) and black cottonwood (93) to strictly anaerobic with no oxygen present in eastern cottonwood (222). Carbon dioxide, nitrogen, and oxygen are the major gaseous components of normal wood, but oxygen was absent or in very reduced amounts in the wetwood studied, indicating that any bacteria present must be either facultative or obligate anaerobes. During the summer growing season, high positive gas pressures have been recorded in wetwood of standing hardwood trees, and these pressures are attributed to bacterial metabolism (33, 79, 191, 221, 222). Trunk gases contributing to the positive pressures in wetwood are carbon dioxide, hydrogen, methane, nitrogen, and hydrogen sulfide. Zeikus and Ward (222) established that the methane gas is produced by an autotropic anaerobe which is a secondary invader and not common to all wetwood populations even within the same host species. This methanogen, subsequently characterized and named Methanobacterium arbophilicum by Zeikus and Henning (221), is also found in soil and water. It can only grow under strictly anaerobic conditions and in the living tree only in wetwood where it metabolizes hydrogen and carbon dioxide produced by Clostridium and other anaerobic bacteria. These other bacteria must, in turn, derive nourishment from the woody tissue.

Van der Kamp et al. (93) consider wetwood with its near anaerobic conditions a perfectly natural phenomenon that imparts resistance to decay to the inner wood of black cottonwood trees.

Strength Properties

There are conflicting reports in the literature concerning the relative strength properties of wetwood compared with normal sapwood and heartwood. Some investigators found wetwood weaker than normal wood; others report wetwood to be equal and even greater in strength. We have found from observations made in conjunction with various tests on sawing, machining, drying, and mechanical strength properties that wetwood is often weaker than normal wood in bonding strength of the compound middle lamella between wood cells. The secondary wall of wetwood cells does not appear to be weaker than that of normal wood with similar specific gravity. Conflicting reports on the comparative strength properties of wetwood can usually be resolved by normalizing test results for sample moisture content, sample size and methods of preparation, and specific gravity of samples.

Results from mechanical tests showing wetwood of hardwoods to be weaker than normal wood were all derived from the testing of green wood (40, 41, 80, 106, 133); furthermore, the wetwood was usually lower in specific gravity than adjacent normal wood. It is important that test specimens be cut from dried wetwood rather than from green wetwood and then dried: during drying, deep surface checks, ring failure, and internal honeycomb checks are much more likely to develop in wetwood than in normal wood. Lagerberg (117) and Thunell (189) found that Scots pine wetwood was weaker in bending, compression, and impact strength properties than were normal sapwood and heartwood. They attributed the lower strength of wetwood to weakness of the middle lamellae and seasoning checks in the air-dried test specimens. Test specimens cut from green wetwood will not have seasoning checks, but green wetwood is more likely to have tissue with weaker bonds between cells than samples selectively cut from the defect-free portion of dried wetwood blanks or boards. Toughness strength of green wetwood samples from white fir (212, 215) and from American and slippery elm (167) was not significantly lower than normal wood when differences in specific gravity were accounted for.

If specimens used in mechanical testing can be prepared from dry, defect-free wetwood, then strength differences between wetwood and normal wood diminish. Stojanov and Enthev (181) found that air-dried specimens from wetwood of silver fir were equal in static bending strength and stronger in compression strength than was normal wood. Table 3 shows that small test beams of wetwood cut from kiln dried boards of California white fir are as strong in static bending properties as is normal wood. Differences between strengths of wetwood and normal wood were related to differences in specific gravity: however, during the preparation of the test specimens, many wetwood samples developed shelling failures and could not be tested. Table 3 also shows that sugar pine wetwood was weaker than normal wood even though higher in specific gravity.

Table 3--Comparison of specific gravity and mechanical strength properties of normal wood and wetwood (sinker heartwood) from California white fir and sugar pine[1]

Species and wood type	Specific gravity oven dry weight, based on volume at test		Static bending strength properties/					
			Modulus of rupture		Modulus of elasticity		Stress at proportional limit	
	Mean	Standard deviation	Mean	Standard deviation	Mean	Standard deviation	Mean	Standard deviation
	12-percent moisture content		Pounds per square inch		Thousand pounds per square inch		Pounds per square inch	
White fir:								
Sapwood	0.353	0.012	9,601	787	1,313	90	5,968	611
Heartwood	.343	.018	9,054	870	1,355	97	5,578	628
Wetwood	.429	.072	11,329	2,347	1,649	338	5,997	1,253
Sugar pine:								
Sapwood	.266	.016	5,643	587	708	106	3,588	430
Heartwood	.271	.024	6,049	702	686	87	3,883	491
Wetwood	.281	.020	5,529	606	674	68	3,041	369

1/Test samples from kiln-dried lumber. Unpublished data from J. C. Ward, U.S. Forest Products Laboratory, Madison, Wis.
2/Strength values at 12-percent moisture content from test specimens measuring 1 by 1 inch in cross-section and 16 inches along the grain (bending test over a 14-inch span).

The intrinsic weakness of wetwood in cell-to-cell bonding strength appears to be related to the production of pectin-degrading enzymes by bacteria, especially the obligate anaerobe Clostridium. These enzymes degrade the pectic substances in the compound middle lamella that holds together the wood cells. Wong and Preece (219) observed degradation of the middle lamellae of cricket-bat willow by the facultative anaerobe Erwinia salicis which also produces pectolytic enzymes. Soluble sugars could not be detected in the wetwood of Abies alba, so Bauch et al. (20) assumed that pectinaceous compounds and hemicelluloses serve as substrates for wetwood bacteria. Bacteria isolated from wetwood have not been found capable of degrading lignin and cellulose or the secondary cell wall under laboratory conditions (214, 217). Furthermore, microscopic examination of wetwood indicates that secondary walls are not visibly degraded as are the walls of wood decayed by fungi or attacked& wood-destroying bacteria (79, 106, 107, 117, 122, 130, 158, 164, 213).Hillis et al. (83) consider wetwood bacteria responsible for shake in trees.

Sometimes a strong, pectin-degrading bacterium will be absent from the microbial population in wetwood and shake, deep checks, and honeycomb will not appear in the wood. For example, when American elm wetwood was without Clostridium in its microbial population, the wood did not develop deep surface checks, honeycomb, or ring failure during kiln drying (24). American elm with wetwood infected by Clostridium is highly susceptible to ring failure and shelling when subjected to drying and machining stresses and is referred to as "onion elm" in the lumber trade.

Ordinarily, sapwood will not develop ring shake and honeycomb, but these defects can occur if sapwood is submerged and infected by bacteria that produce pectolytic enzymes (180). Salamon (160) found that honeycomb and shake occurred in dried sapwood boards from dead lodgepole pine and Engelmann spruce trees salvaged from flooded areas. He observed heavy concentrations of bacteria in this sapwood under the electron microscope and considered bacteria responsible for the defects. Boards from sprinkled or water-stored Scots pine and Norway spruce logs were also found by Boutelje and Ihlstedt (26) to be more liable to checking than boards from newly felled trees. A wide variety of bacteria, including anaerobic pectin-degrading Clostridium spp., were isolated from sapwood of water-stored Scots pine by Karnop (95) but not from the heartwood.

Permeability

Wetwood has long been considered to have low permeability because it is extremely slow to dry and wet pockets are often present in dry lumber (79). Several studies have correlated slow drying of lumber with restricted flow of liquids or gases through small cores of wetwood from aspen (98, 106), western hemlock (111, 120, 121), and western redcedar (188). On the other hand, the absorptive capacity of wetwood can be greater than that of normal wood. In the green condition, white fir wetwood is slow drying, but it will absorb more oil than normal wood does after drying (9, 215, 217). Air-dried wetwood from Scots pine has greater water absorption capacity than normal wood and the capillary rise of water is much more rapid, yet the lumber will contain wet pockets with moisture content as high as 100 percent (117). The moisture-holding capacity of elm wetwood does not exceed that of normal wood (167), but elm lumber with wetwood does not dry abnormally slowly either (24).

For a better understanding of the total concept of permeability for wetwood, studies should be designed to evaluate the relative contributions of wetwood extractive, pit aspiration, tyloses, and bacterial metabolism. Wetwood can have concentrations of extractives that not only may block cell lumens and pits, but can also have greater osmotic pressures than normal wood (6, 20, 45, 83, 106, 112, 117, 136, 139, 165, 169). Reduced permeability of wetwood has been associated with aspiration of bordered pits in softwoods (111, 121, 203) and tyloses in hardwoods (98, 106, 205). Bacteria have produced slime or extracellular polysaccharides in the wetwood of both conifers and hardwoods (158, 203, 205). Bacterial slime can contribute to blockage of cells and increased osmotic pressure.

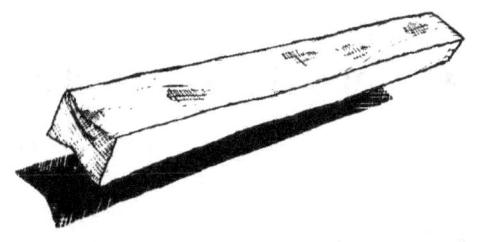

In contrast to wetwood, the permeability of water-stored logs is increased, rather than decreased, by bacteria. Porosity of water-stored wood is increased by bacterial degradation and destruction of pit membranes and thin-walled parenchyma cells, especially in the rays (21, 26, 50, 55, 63, 72, 90, 94, 95, 105, 119, 157, 195, 208). To a limited degree, cell walls in water-stored sapwood may be pitted and corroded (44, 69, 70, 72, 76, 119), suggesting that bacteria may be attacking the wood under aerobic rather than anaerobic conditions. Karnop (94) found that the anaerobic *Clostridium omelianski* can attack unlignified cellulose in the sapwood of water-stored pine, but this activity is reduced with low water temperature and acidic conditions.

Chemical Brown Stain Precursors

Dark discolorations or chemical brown stains frequently develop on the surface of wetwood which, after being removed from the anaerobic atmosphere of the tree, is exposed to aerobic or oxidative conditions. The intensity of these chemicals or oxidative brown stains varies with wood drying conditions, tree species, and individual trees. Wetwood has been associated with dark discolorations in both hardwoods and softwoods (79), particularly in these species: aspen (79, 84, 106, 133), white popular (71), eastern white pine (35), western redcedar (136), redwood (5), western hemlock (18, 165), "Sugi" (*Cryptomeria japonica* (L.F.) D. Don) (66), paper birch and sugar maple (79), and cricket-bat willow (47, 209, 219).

Bacterial degradation of the normal wood sugars and polyphenols may cause wetwood to develop dark oxidation stains when dried. Chemical brown stain has been attributed to the bacterial attack on extractives in sugar pine (182), western hemlock (61), and Pacific silver fir (19). The Forest Products Laboratory's (FPL) investigations (see footnote 8) found that the formation of chemical brown stain in freshly sawed green lumber from California white fir, sugar pine, eastern white pine, western hemlock, aspen, and cottonwood is associated with and confined to the wetwood zones of these trees. This stain, however, did not develop uniformly in all wetwood of these species; variations in composition of bacterial populations appear to be important. FPL data suggest that the presence of phenyloxidizing bacteria similar to those described by Greaves (70) is required before very dark chemical brown stain develops in wetwood.

Chemical brown stain in wetwood is identical to the so-called sour log brown stain that develops on the surface of softwood lumber sawn from logs stored under water or in moist, shaded log decks. Millett (141) observed that brown stain precursors can be present in sugar pine trees, or the precursors can form by enzymatic or hydrolytic degradation in stored logs. Both types of brown stain may have similar biochemical origins. Bacteria are considered factors in the formation of sour log brown stain on sapwood of western pines (55, 182, 183). Knuth (104) discovered that chemical brown stain, but not decay, will develop on the surface of wood samples submerged in liquid cultures of bacteria. Formation of stain was enhanced by autoclaving the wood before inoculation, and aerobic conditions were necessary for final development of stain. Evans and Halvorson (61) studied the chemistry of brown stain in water-stored sapwood and wetwood of western hemlock. From their research, they were able to postulate that bacterial enzymes (probably polyphenol oxidases) condense monomeric leucoanthocyanins to water soluble polymers. These polymers migrate to the wood surface during drying and on exposure to the atmosphere undergo oxidative condensation to dark brown polymers.

The presence of wetwood can cause substantial economic losses when the affected timber is converted into logs and end-use products. Since wetwood is usually limited to inner sapwood and heartwood, it seemingly is not detrimental to growth of trees. In trees where wetwood radiates into the outer sapwood from the central core because of insect attack and other stem injuries, the function of sapwood may be impaired. To what extent this may affect the growth of the tree has not been investigated.

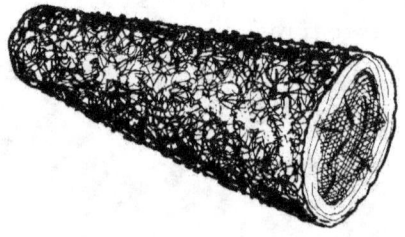

Shake and Frost Cracks

The weaker bonding between the cells of wetwood can result in radial and tangential growth ring separations when the affected trees are subjected to stresses from wind, growth, and freezing. These separations, known as shake (radial and ring) and sometimes spangles, are internal defects of the trunk that often go undetected when the tree appears vigorous and healthy. Shake can make boards with otherwise sound and clear wood, worthless for structural and factory grade lumber. The construction industry is the largest wood-using industry, and the losses in available construction grade lumber resulting from shake in both trees and boards are considerable.

Many logs, both hardwood and conifer, are excluded from the most valuable log grades for lumber and veneer only because they contain shake (81, 117, 124, 125, 126, 127, 128). Shake in western hemlock logs from Alaska has caused staggering losses in the volume of wood intended for export to Japan.[15/] An association between shake and wetwood has been proposed by many investigators (31, 33, 37, 79, 83, 96, 110, 111, 117, 202, 203, 204). Even though wetwood is present in conjunction with much of the shake observed in trees and logs, the actual formation of shake is a complex process involving other factors, such as tree growth stresses, stem injuries, and anomalous wood tissue (38, 83, 91, 113, 114, 115, 124, 129, 130, 140, 172, 173).

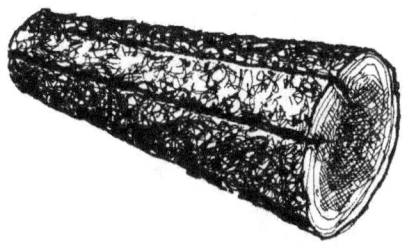

Frost cracks on the outer stem surface appear to be an extension of shake from within the bole; they result not only in loss of usable wood volume but also in a reduced yield of high quality lumber or veneer from the logs. Wood from trees with frost cracks is likely to contain wetwood (see footnote 3) (46, 79, 88, 117). In European poplar plantations, frost crack is a common and serious defect consistently associated with bacterial wetwood in affected trees (57, 62, 82). Surveys of defect in western conifers do not indicate a close relationship between frost cracks and fungal decay (27, 65), yet frost cracks certainly offer a favorable infection court for decay fungi. Aho (2) found that less than 6 percent of the frost cracks in grand fir from Oregon and Washington were infected by fungi, and decay losses were small. He did find that all grand fir trees with frost cracks invariably had wetwood, and 46.7 percent of all trees 11.0 inches or more in diameter contained frost cracks.[16/] From an extensive review of the literature, Schirp (163) found that the association of frost cracks with wetwood in tree stems dated back to 1765.

[15/]R. O. Woodfin, Pacific Northwest Forest and Range Experiment Station, Portland, Oreg., personal communication to J. C. Ward.

[16/]P. E. Aho, Forestry Sciences Laboratory, Corvallis, Oreg., unpublished data.

Rapidly freezing air temperatures are also necessary for formation of frost cracks. In Macedonia, frost cracks developed in the stems of poplars (Populus x euramericana (Dode) Guinier) growing in an area with a continental climate (warm summer temperatures and cold oscillating winter temperatures), but not a single stem crack was observed on trees growing in an area with a Mediterranean climate (mild winters and hot summers) (74). Robert Hartig (77), the "father" of forest pathology, observed frost cracks to be most abundant on the northeast side of trees where sudden and large drops in temperature often occurred in winter. He also noted radial and peripheral cracks in the interior of old oaks; these cracks did not extend outside the stem, and he was uncertain whether they were due to frost. A survey of damage from frost cracks on walnut (Juglans regia L.) growing in the Ukraine showed that the number of stems affected ranged from 1.1 to 83.4 percent in stands on south slopes with fairly moist soils (185). White fir stands of the Sierra Nevada in California have two to three times more frost cracks in trees on the east slope with greater drops in temperature than on the milder west slope (196). Freezing of redwood boards results in breakage of heartwood with wetwood but not with normal heartwood below 150-percent moisture content (58).

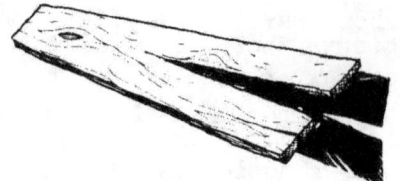

Losses in product volume and value associated with shake and frost cracks are especially important because these defects are most prevalent in the lower two logs of trees and, in many instances, are spiraled (see footnote 3) (163, 196). In the true firs, for example, these logs have the potential for producing not only the highest quality boards but also nearly half the total lumber volume of the tree (215). It is in these logs that reducing wetwood-related losses has its greatest potential, both in volume and value (147). The occurrence of combinations of shake, frost crack, and spiral grain in white spruce timber growing in northern Alberta seriously affected the yield of 1- and 2-inch dimension lumber (131). Shake caused 7 to 12 percent of the lumber output per tree to be graded utility or lower. Degrade was higher, 15 to 22 percent, when frost cracks were present; and trees with both frost cracks and spiral grain suffered the highest grade losses--23 to 26 percent.

Checking and Collapse

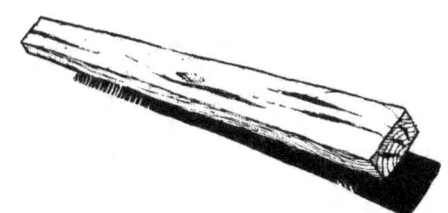

The unexpected occurrence of checking and collapse in lumber and veneer during drying can be attributed to the same weakening of the compound middle lamella that predisposes the wood to shake and formation of frost cracks in trees. In wetwood, drying checks can develop in both a radial and a tangential direction. Radial checks may be deep surface checks, internal ruptures called honeycomb, or bottleneck checks-- deep surface checks that develop into a honeycomb. Tangential checks known as ring failure appear to be an incipient form of ring shake that starts in the tree: these checks do not actually rupture until subjected to shrinkage stresses (202). A description by Kutsche and Ethington (116) of various shelling failures during the machining of wood suggests that some failures may be related to wetwood and may possibly be an incipient form of ring shake. Collapse is essentially a collective internal failure of cell walls resulting in depressions on the surface of the dried board or veneer which cannot always be removed with surfacing. Figures 1 and 2 show examples of honeycomb, ring failure, and collapse that developed in wetwood during drying. Deep surface checks are present in some pieces but are not visible because they close up toward the end of drying when the surface is below 25-percent moisture content (MC) while the core is still above 30-percent MC.

The dry kiln operation is often blamed for shake in dried lumber. Results from investigations with softwood dimension lumber reveal, however, that much of the observed shake was initiated in the tree and only became apparent after drying (37, 87, 140, 203). Shake in western hemlock lumber has been associated more with wetwood than with drying conditions (111, 203).

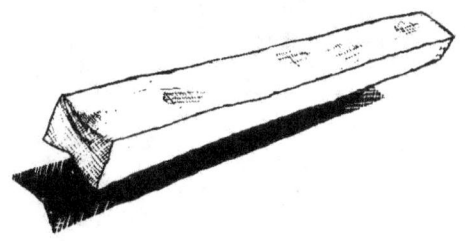

Green lumber with wetwood is more likely to develop collapse during the early stages of drying than lumber with normal sapwood and heartwood. Collapse can be expected in wetwood when the bonding strength between cells has been weakened and the rate of internal moisture loss through cell cavities is restricted. A generally held theory is that collapse can develop only in cell cavities that are completely saturated with water (142), but Kemp (98) was able to induce collapse in wetwood cells of aspen that were not completely saturated. He was also able to relate collapse to portions of the board with the lowest rates of moisture loss. Honeycomb and ring failure are often associated with collapse in wetwood. Reconditioning of collapsed lumber by steaming is possible when internal ruptures are not present (108).

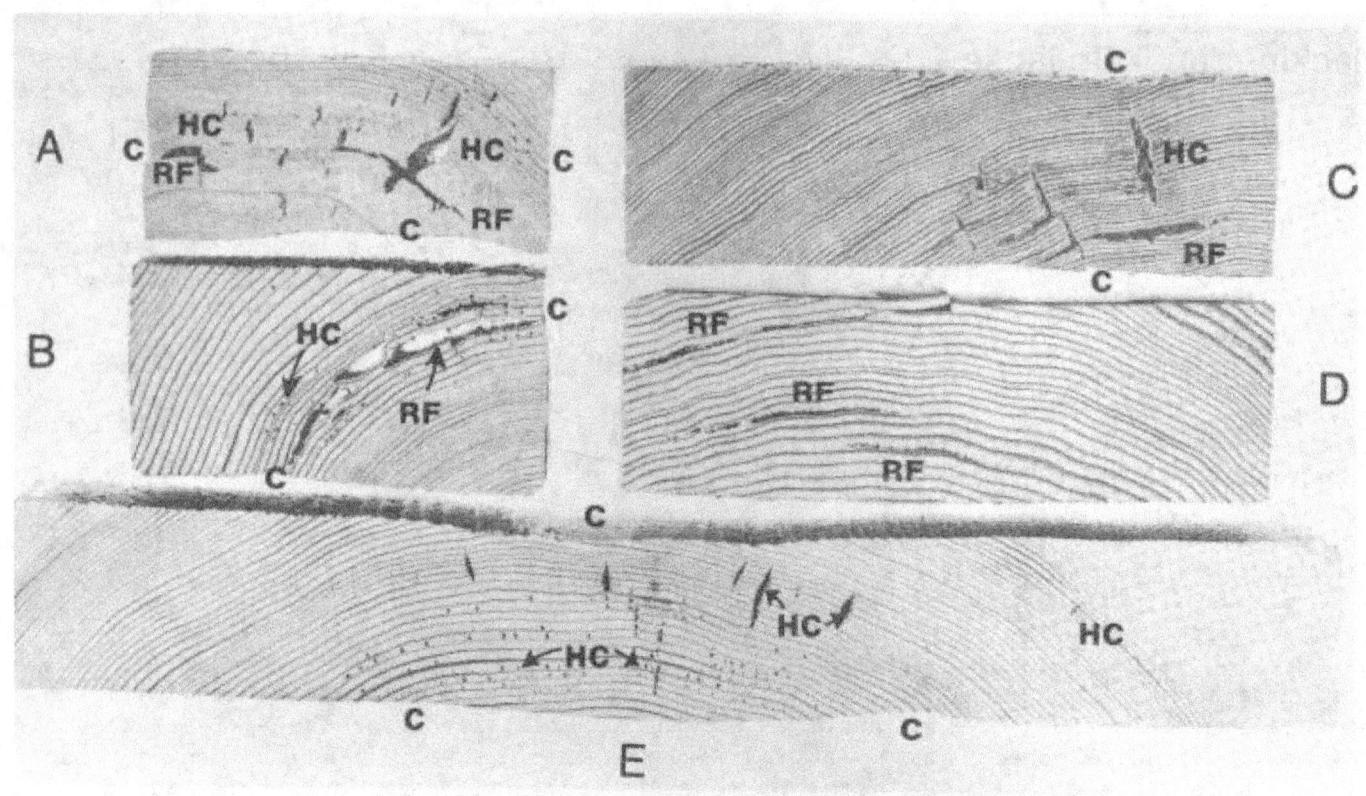

Figure 1 .--Cross-cut sections from softwood boards with wetwood that developed collapse (C), honeycomb (HC), and ring failure (RF) in drying: **A**, old-growth eastern hemlock; **B**, young-growth eastern hemlock; **C**, old-growth western hemlock; **D**, young-growth western hemlock; **E**, California white fir.

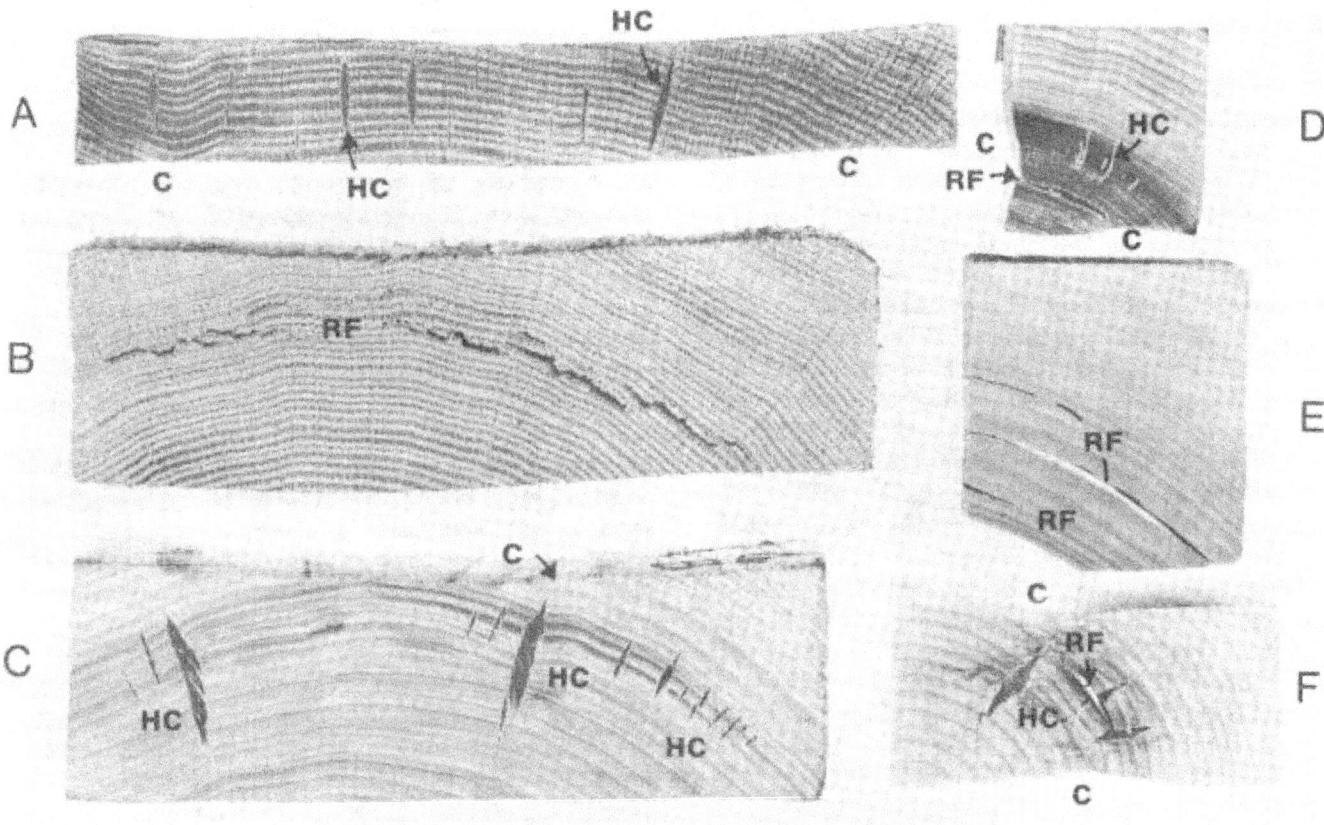

Figure 2. --Cross-cut sections from hardwood boards with wetwood that developed collapse (C), honeycomb (HC), and ring failure (RF) in drying: **A** and **B**, northern red oak; **C**, eastern cottonwood; **D**, red maple; **E**, American sycamore; **F**, largetooth aspen.

Two major sources of stress considered responsible for collapse during drying are liquid water tension produced by capillary forces in the cell which can be very high with rapid drying and compression stress perpendicular to the grain caused by a sharp moisture content gradient across the board (108, 142, 151). Little information is available on the comparative susceptibility of wetwood and normal wood to collapse. Kemp (98) found that aspen heartwood, artificially saturated with water, will collapse but only at a higher temperature than that required to first initiate collapse in wetwood. High drying temperatures can plasticize wood and make it less able to withstand the stresses associated with collapse (108). Thin-walled cells in the newly formed earlywood of outer sapwood have collapsed when the living tree is subjected to extreme moisture stress (198). Results from drying tests at the Forest Products Laboratory (see footnote 8) indicate that wetwood is susceptible to collapse under kiln-drying temperatures less than 212°F, but normal wood is not.

Collapse, honeycomb, and ring failure are costly defects associated with the drying of wetwood from white fir[17] (145), western hemlock (100), redwood (138), western redcedar (51, 73, 188), aspen (41, 98, 133, 134) ,and west coast hardwoods (176)--particularly Pacific madrone (30), tanoak (154), and California black oak (205). The plugging of cell cavities with tyloses increases the tendency of wetwood to collapse in such hardwoods as aspen (98), overcup oak (126), California black oak (205), and red gum (151).

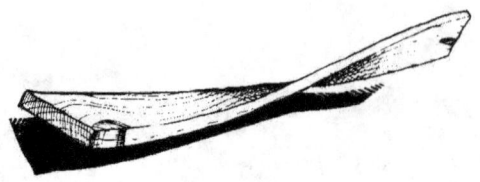

The severity of honeycomb and collapse in wetwood usually increases with an increase in temperatures during the early and middle stages of drying. Temperatures at which these defects begin to form in wetwood apparently vary with tree species, site conditions, and types of wetwood. When green lumber is kiln dried, honeycomb in the wetwood of California black oak and northern red oak can be minimized by lower initial dry bulb temperatures (202, 205). Wetwood in southern bottom-land oaks, however, honeycomb and collapse under the mildest kiln schedules; this lumber must initially be dried in the open air or dried in kilns at low temperatures to at least 25 percent MC before it can be kiln dried.[18] Collapse in aspen wetwood can be reduced by using mild kiln conditions in the early stages of drying (98). Western redcedar lumber containing a particularly dark heavy type of wetwood will sometimes collapse after only 1 day of air drying.[19] With high temperature kiln schedules (temperature of 212°F or higher) most wetwood will honeycomb or collapse, but usually adjacent normal wood will not[20] (36, 132, 134, 159).

_____

[18]/Don Cuppett, Northeastern Forest Experiment Station, Princeton, w. Va., personal communication to J. C. Ward.

[19]/Jack McChesney, Louisiana Pacific Corp., Dillard, Oreg., personal communication to J. C. Ward.

[20]/Unpublished data from drying studies: on file at U.S. Forest Products Laboratory, Madison, Wis.

_____

[17]/B. G. Anderson. Study of change in grade and footage loss in kiln dried white fir lumber from the green chain to the car. Report presented at the meeting of Eastern Oregon-Southern Idaho Dry Kiln Club, La Grande, Oreg., Nov. 19, 1954. 4 p. Oregon Forest Products Laboratory, Corvallis, Oreg.

Honeycomb and ring failure are major defects associated with kiln drying of oak, our most important hardwood lumber species. Annual wood losses from drying defects associated with wetwood in oak factory grade lumber produced in the Eastern United States are estimated to be at least 3 percent. It is not unusual, however, for 10 to 25 percent or even half the lumber charges from individual kilns to be lost because of honeycomb and ring failure.

FPL studies on kiln drying northern red oak lumber, green from the saw, showed 8- to 25-percent losses in volume from honeycomb and ring failure in wetwood (201). Monetary losses ranged from $34.28 to $139.42 per thousand board feet of rough dry lumber. Three studies of kiln drying California black oak (205) at a commercial mill showed losses from defects in wetwood to be 7 to 48 percent of the total volume of Number 1 Common and Better lumber. Total monetary loss in rough, dry lumber for the three studies was over $9,000. This can explain why one large Los Angeles lumber company fills a standing monthly order for one-half-million board feet of 4/4-inch kiln-dried oak lumber with oak imported from the Eastern United States.

Smith (180) estimates that, for British Columbia, the development of honeycomb and ring shake in softwood studs with bacterially infected wood will result in a reduction in grade of $7 per thousand board feet. We can calculate from the average grade prices in "Random Lengths" (150) that drying defects associated with wetwood can cause grade losses of $38 to $40 per thousand board feet for hem-fir dimension lumber.

Slow Drying Rates and Uneven Moisture Content

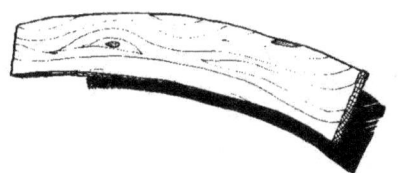

The wetwood of many species has low permeability and requires much longer drying times than normal wood to reach a desired moisture content. Even when wetwood boards reach the desired average MC, there is an uneven distribution of moisture where the shell is very dry but the core contains wet pockets or streaks that are still above the fiber saturation point. Sometimes after supposedly dry wetwood lumber is surfaced, the internal wet pockets dry and collapse, resulting in degrade of the lumber (134). Wet pockets can be a problem when dried wetwood from aspen and hemlock are used for core stock in the manufacture of doors and panels. Although these wet pockets may be pencil thin, they will build up enough steam pressure during electronic gluing operations to explode and shatter the surface of the pieces.21/ Internal wet pockets in kiln-dried western hemlock causes erroneous determinations of MC when electronic meters are used (109).

21/ Elmer Cermak, Algoma Hardwoods, Inc., Algoma, Wis., personal communication to J. C. Ward.

A serious economic problem can result from the presence of impermeable wetwood in softwood dimension lumber for construction purposes. The same problem applies to aspen and poplar studs and light framing lumber graded under softwood rules. To meet moisture specifications for kiln-dried lumber softwood dimension lumber must be dried to at least 19-percent MC; the desirable range is 12 to 16 percent. Boards with wetwood may require 50 percent or more time in the dry kiln to reach 19-percent MC as do boards with sapwood and heartwood. When the volume of boards exceeding the 19-percent MC specification is greater than 5 percent, a kiln charge or shipment of kiln-dried lumber is considered to be in noncompliance with specifications for moisture[22]/ (28, 178, 210, 218).

Slow drying of construction lumber containing wetwood can be expected for these softwood species: redwood (103, 138, 151), true firs (101, 103, 145, 151; also see footnote 17); western hemlock (29, 103, 110, 151), western redcedar (29, 103, 151, 159), and white pines (151). Hardwood species with impermeable wetwood and slow drying rates are: aspen (36, 86, 133, 200), red gum (151), and water and swamp tupelo (137). Redwood with heavy sinker heart is especially difficult to dry (138), and 1-inch-thick boards require 146 days of air drying to reach 20 percent MC (52). In contrast, 2-inch-thick redwood with light and medium heartwood can be air dried to 19-percent MC in 74 days (8). Arganbright and Dost (7) found that development of chemical brown stain decreases the drying rate of sugar pine and, at the end of drying, retards moisture movement back into the board surface during conditioning treatment.

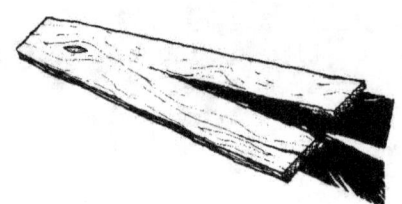

Table 4 shows that the slow drying rates of wetwood vary with tree species. Wetwood from aspen and the conifers is much less permeable than normal wood, whereas wetwood from cottonwood, elm, and oak takes only slightly longer to dry than normal wood. The effect of wetwood on reducing drying rates tends to decrease as thickness of boards decreases. For some species, the time required to dry veneer containing wetwood and veneer containing sapwood differs little for a given MC: but for species such as red gum, true firs, hemlock, larch, and redwood, there are decided differences (126). A comparison of drying times for green veneer from balsam fir and white fir (table 5) indicates that wetwood has an adverse effect on drying rates for even thin material.

Where wetwood boards occur in kiln charges of softwood and aspen dimension lumber, the processor is confronted with five possible alternatives; each may result in losses of wood and energy and in higher manufacturing costs.

1. The kiln residence time can be extended until the wet test boards reach the required 19-percent MC. This can cause overdrying of nonwetwood stock (i.e., below 11-percent MC), which is then subject to planer splits and costly degrade in subsequent machining operations (100, 101, 103, 110, 145, 218; also see footnotes 17 and 22).

22/G. Reinking. 1971. Kiln drying lumber to the moisture provisions of the new lumber standards. Forest Products Research Society News Digest File J-1.1, 2 p. Madison, Wis.

Table 4--Drying times for green wetwood boards compared with drying times for boards containing normal wood[1]

Species	Source of data	Board specifications		Initial kiln temperatures [2]		Wetwood		Sapwood		Heartwood	
		Thickness	Dried moisture content	DB	WB	Green moisture content	Drying time	Green moisture content	Drying time	Green moisture content	Drying time
		Inches	Percent	- - °F - -		Percent	Hours	Percent	Hours	Percent	Hours
Softwoods:											
Western hemlock	Ward and Kozlik (203)	1-34	15	180	176	150	139	156	82	54	59
Western hemlock	Kozlik et al. (111)	3/4	10	100	69	153	170	--	--	66.1	90
Eastern hemlock	J. C. Ward[3]	1-3/4	15	180	170	148	185	137	95	75	96
White fir	Smith and Dittman (177)	1-7/8	20	160	145	193	158	145	84	57	42
White fir	Smith and Dittman (177)	1-7/8	4/15	160	145	193	4/195	145	4/97	57	4/63
White fir	Pong and Wilcox (147)	1- 1-3/4	15-16	160	140	--	174	--	106	--	60
White fir	J. C. Ward[3]	1-3/4	15	180	170	155	150	170	105	66	73
Sugar pine	J. C. Ward[3]	1-1/2	15	120	5/75	194	226	204	148	51	72
Hardwoods: [6]											
Aspen	Cech (36), Huffman (86)	1-3/4	15	140	133	170	4/210+	140	175	--	--
Quaking aspen	J. C. Ward (200)	1-3/4	15	180	170	115-128	179	76-122	90	81-106	115
Eastern cottonwood	J. C. Ward[3]	1-1/8	7	180	170	144	75	131	85	--	--
American elm	J. C. Ward[3]	1-1/8	7	120	113	106	134	90	121	69	113
California black oak	Ward and Shedd (205)	1-1/8	7	120	115	95	387	--	--	85	379
Northern red oak	J. C. Ward[3]	1-1/8	7	125	121	97	324	--	--	86	312

[1]/Comparisons are within kiln runs and not among kiln runs or between species.
[2]/DB = dry bulb temperature; WB = wet bulb temperature.
[3]/J. C. Ward, unpublished data on file at U.S. Forest Products Laooratory, Madison, Wis.
[4]/Extrapolated from published data.
[5]/Kiln vents open with steam sprays off.
[6]/All wetwood developed honeycomb, collapse, or ring failure or a combination of these.

Table 5--Drying times for green veneer with wetwood compared with veneer containing normal sapwood and heartwood

Species	Source of data	Veneer specifications		Dryer temperatures	Wetwood		Sapwood		Heartwood	
		Thickness	Dried moisture content		Green moisture content[1]	Drying time	Green moisture content[1]	Drying time	Green moisture content[1]	Drying time
		Inch	Percent	°F	Percent	Minutes	Percent	Minutes	Percent	Minutes
Balsam fir	Dokken and Lefebvre (48)	1/10	4	300	154	18-1/2	201	15-1/4	76	11
	Dokken and Lefebvre (48)		4	450	154	7-3/4	201	7-1/4	76	5-1/2
	Dokken and Lefebvre (48)	1/6	4	300	154	35	201	31	76	25-1/2
	Dokken and Lefebvre (48)		4	450	154	15	201	14-1/4	76	9-1/2
White fir	Vern Parker[2]	1/8	5	3/	150	10	150	8		4/

[1]/ Average value.
[2]/ Superintendent, Plywood Mill, Bendix Corp., Martell, Calif., personal communication to W. Y. Pong.
[3]/ 4-stage gas jet-dried: 450°, 450°, 400°, and 350°F.
[4]/ Heartwood dried according to sapwood schedule of 8 minutes will be overdried with a moisture content of 1 percent.

2. The wet stock can be sorted and redried after a normal kiln run. This approach will increase drying and handling costs for the wetwood lumber. For white fir dimension lumber containing wetwood, redrying can result in a 40-percent increase in the kiln-drying costs.[23]

3. Boards greater than 19-percent MC can be marketed as surfaced green lumber and less return will be received for the product. For hem-fir dimension lumber, this can amount to approximately $26 per thousand board feet in selling price ([149]), not to mention drying and extra handling losses. Shipping weights and the resulting shipping costs of green lumber are important considerations in this third alternative. In table 6 are calculated weights for 1,000 board feet of white fir at different moisture contents. Also included are costs for shipping this lumber to Los Angeles and Chicago from Portland, Oregon. The price differential of $26 between green and dry lumber barely offsets the shipping cost differential of dry lumber (15 percent) and green lumber (150 percent) to Los Angeles but not to Chicago.

---

[23]Donald O. Prielipp, general manager, Roseburg Lumber Co., Anderson Division, Anderson, Calif., personal communication to J. C. Ward.

Table 6--Freight costs for shipping white fir lumber[1]

Moisture content	Calculated weight per 1,000 board feet	Cost of shipping 1,000 board feet	
		To Los Angeles (base rate, $1.27/100 pounds)	To Chicago (base rate, $3.07/100 pounds)
Percent	Pounds	- - - - - - - - Dollars - - - - - - - -	
0	1,271	16.14	39.02
12	1,423	18.07	43.69
15	1,463	18.58	44.91
19	1,513	19.22	46.45
25	1,590	20.19	48.81
75	2,226	28.27	68.34
100	2,543	32.30	78.07
115	2,734	34.72	83.93
150	3,178	40.36	97.56

[1]F.o.b. Portland; base rates (Dec. 15, 1978) are for railroad shipments of 85,000 pounds or less (Western Wood Products Association, Portland, Oreg.).

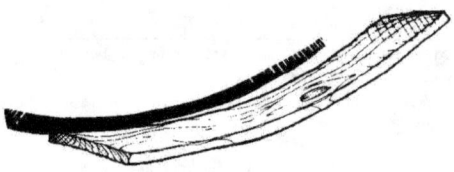

4. The lumber can be dried by a normal schedule and not sorted for wet stock. All boards, both above and below 19-percent MC, would be surfaced and sold as kiln dried. This option could be exceedingly costly if the volume of wet stock exceeded the 5-percent limitation ([210]) and was detected by the buyer and reinspection by the grading association requested. Not only would the surfaced lumber have to be resorted, regraded, and metered for moisture, but the boards that exceeded the maximum allowable moisture would be considered substandard surfaced green and regraded accordingly. Losses in excess of $50 per thousand board feet are possible with reinspection.

5. The green lumber could be segregated into uniform drying sorts (sapwood, heartwood, and wetwood or sinker) and each then dried under different kiln schedules or treatments. This presorting would minimize the extremes in final moisture content within a given kiln charge and has long been advocated for western softwoods (103, 178), particularly white fir[24] (101, 177, 179; also see footnote 17), western hemlock (110), redwood (138), and incense-cedar (152). Presorting would minimize energy costs. With the energy crisis, it becomes imperative that costs of energy used in drying and redrying wetwood be given careful consideration, especially since 60 to 70 percent of the total energy used in manufacturing most wood products is consumed in drying (43).

Commercial presorting of green lumber for drying by visual detection and hand methods has been used with some success for western hemlock (110), incense-cedar (152), and California white fir (145, 178, 215). For presorting to be effective, however, each board must be examined on all sides during the segregation operation. Under high production mill conditions, manual presorting is generally not possible. Examination of individual boards will also be limited where tray and drop sorters are used. Mechanical, electrical, and optical instruments for automatic presorting of wetwood lumber on a commercial scale have not been developed (207).

Another obstacle to obtaining maximum effectiveness in presorting is the mixture of wetwood and normal wood that occurs in many boards on the commercial green chain (110, 147). It is possible to develop optimum schedules for drying boards having mixtures of normal sapwood and heartwood with wide differences in green moisture content (161), but not for mixtures of sapwood and wetwood even though there may be little or no difference in green moisture content (162). Studies of white fir indicate that boards containing mixtures of wetwood, sapwood, and corky heartwood will have more drying and surfacing problems than boards that dry more uniformly (147, 215).

Steaming of green lumber before drying increased the drying rate of wetwood from white fir (175) and redwood (52), but was not effective with western hemlock (110) or eastern hemlock (see footnote 8). The chemical nature of extractives in the wetwood seems to be an important factor controlling the effectiveness of presteaming.

---

[24] J. Steel. 1953. Kiln drying white fir. Wood drying committee news digest. (Aug.) 3 p. Forest Products Research Society, Madison, Wis.

Chemical Brown Stain

Chemical brown stain in wetwood is an especially serious defect in lumber, veneer, and wood fiber products graded on appearance. Brown stain also affects the marketability of softwood dimension lumber because consumers think the wood looks decayed. Wood decay fungi are not considered causal agents for chemical brown stain (51, 56, 85, 103, 148, 151). Brown stain in wetwood usually develops in the zone between sapwood and heartwood or within bacterially infected heartwood. Sapwood of many species develops brown stain if taken from logs stored for an extended time under humid conditions. Brown stain developed in eastern white pine lumber from logs stored in the woods for 42 and 93 days, but not in lumber sawn and seasoned within 24 hours of felling (14). Depth of stain in boards increased as storage time increased. Extended storage of ponderosa pine and sugar pine logs under water sprays will initiate development of brown stain in sapwood boards and further intensify darkening in wetwood zones.25/

Timber species noted for developing chemical brown stain during drying usually tend to have wetwood. Processing problems associated with chemical brown stain in lumber are considered important commercially for the following species: sugar pine, eastern and western white pines, ponderosa pine, Pacific silver fir, noble fir, western hemlock, redwood, western redcedar, and Sitka spruce (5, 7, 18, 19, 23, 29, 35, 56, 61, 89, 99, 103, 141, 151, 165).

Brown stain can develop in both air-dried and kiln-dried wood, but it is usually more pronounced under warm, humid kiln schedules. Considerable brown stain can occur in boards air dried at temperatures as low as 80°F (141). Chemical brown stain may also develop just below, but not on the surface of, air-dried lumber and will not be noticed until after the boards are surfaced (155). This concealed brown stain, referred to as yard brown stain, has been observed in ponderosa pine, sugar pine, and the white pines (85). Wetwood prone to developing brown stain should be segregated from other green lumber. The wetwood stock would be air dried; the normal stock, kiln dried. For minimum development of brown stain during air drying, the lumber should be from freshly felled trees and drying should be fast, with relative humidities below 65 percent (14, 61, 66, 141, 155).

If faster drying of wetwood is desired or presorting is not feasible, reduction or prevention of chemical brown stain during kiln drying may be accomplished by either manipulating kiln schedules or dipping the green boards in enzyme-inhibiting or antioxidant solutions. There are, however, drawbacks to these preventive methods that are important.

25/Del Shedd, Quality Control Supervisor, Kimberly Clark Corp., Anderson, Calif., personal communication to J. C. Ward.

Low initial temperatures in kilns and low humidities with good air circulation are necessary to successfully reduce brown stain or prevent it (23, 29, 99, 102, 105, 141). Rosen (156) found that high temperature jet drying eliminated brown stain in cottonwood, but the wetwood collapsed and checked. With low temperatures, the kiln usually has to be vented to attain low relative humidities (29, 102, 103). Venting dry kilns results in a great waste of energy. Sometimes the kiln doors must be opened during drying to lower the humidity, and the duration of the venting can be from 3 days to 1 week. Manipulating kiln schedules has not been successful in controlling dark stains in redwood unless the lumber was first steamed (5, 52, 54). Solvent drying of redwood with acetone has been proposed for elimination of stains (6).

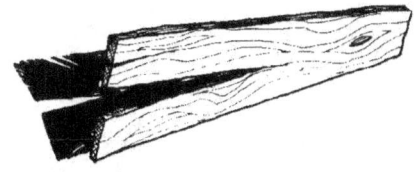

Kiln schedules to prevent stain are not always successful if the drying operation is regulated either on a time basis or on kiln samples containing only normal wood. Figure 3 shows chemical brown stain that developed only in sugar pine boards containing wetwood but not in sapwood or normal heartwood. These boards were dried under an antistain schedule based on the average drying rate of boards with normal wood. The operation could have been based on the drying rate of the wetwood boards, but the normal boards would have been over-dried and the subsequent surfacing of these boards would result in planer splitting. The solution is to segregate the normal boards from the wetwood boards and then dry each board sort under different schedules.

Dipping green lumber in enzyme-inhibiting or antioxidant solutions usually prevents brown stain and permits higher initial kiln temperatures and humidities (34, 35, 61, 102, 103, 170, 182). Such reagents will not reduce brown stain if they are not able to sufficiently penetrate the wood surface during the dip treatment (141). Proper use of antistain dips results in substantial reductions in drying and energy costs, but most chemicals used are highly toxic to humans. Because of Occupational Safety and Health Administration inspections, the dipping of boards in brown stain retardant solutions has been discontinued by many mills.

Stain-producing extractives from wetwood can cause problems with finishes. Figure 4 shows the undesirable darkening of a lacquer finish from an underlying oak laminate containing wetwood. Water-soluble extractives in wetwood of redwood and western redcedar cause serious problems when stains penetrate finishes on exterior siding. Connors found that extraction of sequirins from the redwood can effectively prevent the problem.[26] On a commercial scale this treatment could be cumbersome with lumber, and a disposal problem would be created. Connors also found that dipping the boards in lead acetate before applying finishes is effective but expensive, and lead is now prohibited for such use.

The chemical precursors causing brown stain problems in drying wetwood in western hemlock lumber are also responsible for problems when this species is used for ground woodpulp (15, 17). The cost per ton of raising the brightness of the ground hemlock woodpulp can be increased by 15 to 20 times when brown stain occurs. Aspen wetwood is objectionable for pulp because of added bleaching costs (79).

---

[26]/G. L. Connors. 1968. Considerations for reducing extractives staining in redwood. Office report, 11 p. U.S. Forest Products Laboratory, Madison, Wis.

Figure 3.--Kiln-dried sugar pine shows chemical brown stain in center board that contained wetwood but not in boards with sapwood (left) or normal heartwood (right). End sections (top) were cross cut from rough, dry boards, and one-sixteenth inch was planed from board surfaces (bottom).

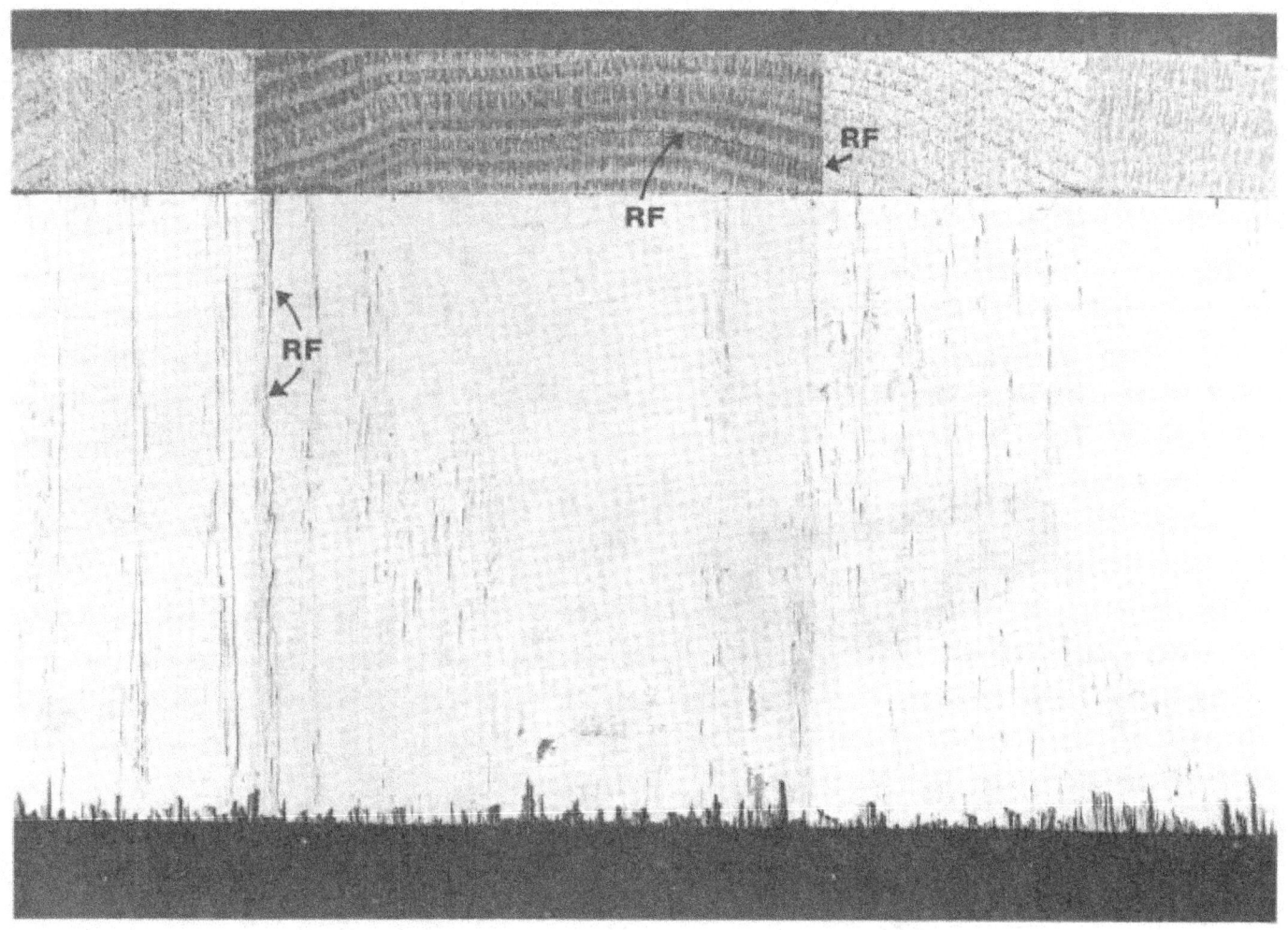

Figure 4.--Darkening of a white lacquer finish overlying a kiln-dried red oak laminate containing bacterial wetwood. Ring failure (RF) within the laminate extended to the surface during air drying of the lacquer finish.

Plywood and Reconstituted Boards

Little is known about the effect of wetwood on the quality of plywood and reconstituted wood products. We believe that investigations of the use of wetwood in these products may well provide insight on unsuspected causes of some processing problems. Also, these products are increasingly promoted for use as components in light frame construction (184), and the effect of wetwood on strength properties should be investigated.

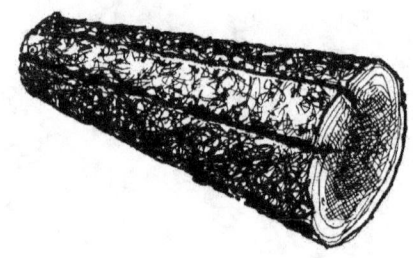

There are several ways that wetwood causes problems with the processing and quality of plywood. Shake and frost cracks cause "spinout" of bolts on the lathe, resulting in splits and splintering of veneer (71, 79, 125, 126). Commercial experience indicates that wetwood in "sinker" logs of species like redwood is undesirable for veneer because of cutting and drying problems (125). The assembly-time tolerance ofconventional plywood adhesion restricts the moisture content of veneer to defined upper and lower limits. Control of moisture content in veneer from wetwood is a major technical problem and can lead to "undercured," "washed out," or "starved" glue lines and to "blows" and "blister" in the pressing operation (32). Nearly all the "blows" during production of white fir plywood are related to wetwood in the veneer.[27]/

The effect of wetwood on the strength of the glue bond is usually associated with wet pockets, and there is a dearth of information on the influence of extractives and pH. Poor adhesion because of uneven distribution of moisture in dried veneers containing wetwood has been reported for eastern cottonwood (191), western hemlock (79), balsam fir (32), and white fir (126). Oxidation stains are objectionable characteristics for face veneers, and these stains may possibly interfere with proper adhesion of the glue to the wood. Some wetwood species most susceptible to oxidation stains during production of plywood are: sugar pine, western white pine, ponderosa pine, Jeffrey pine, red maple, tanoak, bottom-land oaks, tupelo, and willow (126).

A small amount of research data exists to suggest that wetwood may lower the strength properties of reconstituted boards that include particle board and hardboards. Table 7 shows that wetwood from aspen and western larch will not make as strong a flakeboard as normal sapwood and heartwood. Linear stability was not adversely affected by wetwood, but thickness swelling was greater for aspen boards containing wetwood and least for larch. Wet process hardboards from aspen wetwood have a lower modulus of elasticity, modulus of rupture, and internal bond strength than boards from sapwood (67). Dimensional stability of the hardboards from wetwood was greater than boards from sapwood, however.

_____

28/J. C. Ward and J. Chern. 1978. Comparative properties of flakeboard made from normal wood and from bacterial wetwood of trembling aspen and western larch. A preliminary report. 6 p. U.S. Forest Products Laboratory, Madison, Wis.

_____

27/Vern Parker, Bendix Corp., Martell, Calif., personal communication to W. Y. Pong.

Table 7--Physical and mechanical properties of 1/2-inch-thick flakeboards[1] from normal wood and wetwood of trembling aspen and western larch?/

Species and type of wood	Specific gravity?'		Mechanical properties				Swelling in 30- to 90-percent relative humidity	
	Solid wood	Flakeboard	Moisture content	Modulus of rupture	Modulus of elasticity	Internal bond	Linear	Thickness
			Percent	Pounds per square inch	Thousand pounds per square inch	Pounds per square inch	Percent	Percent
Aspen (Wisconsin):								
Sapwood	0.474	0.657	7.2	5,541	692	128	0.07	6.88
Wetwood[4]	.504	.652	7.4	4,499	618	95	.08	9.58
Aspen (Colorado):								
Heartwood	.35	.715	6.2	6,390	994	76	.14	9.56
Wetwood[5]	.34	.634	6.8	4,335	677	48	.14	12.76
Larch:								
Sapwood	.480	.652	9.3	4,992	655	89	.09	11.45
Heartwood	.454	.648	8.9	4,520	635	112	.10	9.87
Wetwood[6]	.701	.628	9.2	3,542	549	74	.08	7.80

[1] Boards made from 0.02- by 2.0-inch flakes with 4-percent phenolic resin.
[2] Data from J. C. Ward and J. Chern. 1978. Comparative properties of flakeboard made from normal wood and from bacterial wetwood of trembling aspen and western larch. A preliminary report. 6 p. U.S. Forest Products Laboratory, Madison, Wis.
[3] Specific gravity of wood based on ovendry weight and ovendry volume. Specific gravity of board based on ovendry weight and volume of board at mechanical test moisture content.
[4] Bacterial wetwood formed where sapwood changes to heartwood.
[5] Bacterial wetwood formed in previously developed, normal heartwood.
[6] Formed from microbially infected inner sapwood--heavily saturated with water-soluble extractives that included arabinogalactans.

Corrosion of Kilns

Reports of accelerated corrosion of dry kilns have increased in frequency in recent years. Most cases of corrosion can be traced to highly acid atmospheres within the kiln during drying. Not only are metal walls, pipes, fasteners, etc., attacked by this corrosive mixture of air, but unprotected concrete (16) and wood stickers are also subject to accelerated deterioration. Several explanations can be given. Drying charges of green lumber rather than partially air-dried lumber considerably increases the liability of wood to evolve corrosive acid vapors (10, 11, 39). Use of high temperature schedules in drying green wood increases corrosion. Contact between dissimilar metals in the kiln, such as aluminum and iron in an acidic atmosphere, will produce a "battery effect" and corrosion of metal will be greatly accelerated (16). The drying of wood treated with aqueous solutions of preservatives will cause corrosion problems (39).

Research Needs

Although investigations are needed, we believe that the presence of wetwood in green lumber charges can be a definite factor in many reported cases of increased kiln corrosion. Oak, hemlock, and the true firs are species likely to have wetwood with a pH as low as 3.5. Metal corrosion is greatly accelerated by a pH of 4.0 or less (10, 11), and acidic vapors of pH 3.5 will attack concrete (16). Barton (16) related the presence of normally occurring chelating compounds in the extractives of western redcedar with the corrosion of iron but not of aluminum in dry kilns. Not all charges of western redcedar caused chelation corrosion.

Wetwood in western hemlock lumber may be an important factor in kiln corrosion. In 1951, MacLean and Gardner (135) reported that an increase in the deterioration of wooden dry kilns was due to increased use for drying western hemlock lumber. An acid condensate of aqueous vapors from the hemlock resulted in deterioration of the cellulose in kiln boards and corrosion of metal fittings. Although Douglas-fir heartwood is acid, the vapor condensates were not as acid as those from hemlock. Also, Douglas-fir heartwood has a higher content of resin, and the vapor condensates form a protective layer. The acid vapors from drying one or two charges of hemlock will remove this protective layer of Douglas-fir condensates. Most lumber is now dried in metal kilns, and Barton (16) estimates that at least 90 percent of the corrosion in these kilns is caused by vapor condensates.

Wetwood is responsible for substantial losses of wood, energy, and production expenditures in timber-using industries. There is a clear need to more thoroughly assess these losses and to find ways of eliminating or minimizing them.

An effective research program on wetwood should include planning short- and long-term solutions to the overall problem and providing for them. Solid wood products, particularly lumber and veneer, are most adversely affected by processing defects related to wetwood. Lumber and veneer are expected to be prime economic products from future timber harvests (97). In the long term, the best solution to the wetwood problem may well result from timber management research to prevent future occurrences of wetwood.

Short-Term

The initial research effort must be primarily concerned with attacking existing utilization and processing problems, but it should also provide a scientific basis for long-term research planning. A short-term research program should concentrate on: (1) assessing the occurrence of wetwood in standing timber and its effect on product yields and losses, (2) defining the chemical and physical properties of wetwood as a basis for detecting and segregating affected wood products, and (3) developing optimum utilization and processing methods for timber and wood products containing wetwood. Item 1 would be the pivotal area of endeavor for the short-term research program and also the area most likely to provide continuity with long-term research.

These studies should be carried out cooperatively by timber managers, timber processors, and forestry researchers. The managers would provide the timber and the mill operators the processing facilities. Researchers would record information on timber quality for selected sample trees. Bucked logs from the sample trees would be followed individually through the milling process and information, including the grade and amount of product cut from each log, tallied. The yield data would be compiled and analyzed in conjunction with timber quality data to provide the basis for predicting the product yield from similar timber. Measurements of wetwood in the timber and the resultant defects in both green and dried wood products would be incorporated into the beginning and final phases of a product yield study.

Studies on timber products and yields are important to the initiation of research on wetwood. Now we can only guess, but with some confidence, that the utilization and processing problems associated with wetwood are costing the wood-using industries many millions of dollars annually. No statistical or economic surveys have been initiated either for the purpose of determining the amount of wetwood in timber stands or for estimating the losses that unquestionably result from the processing of wetwood timber. Government agencies have not initiated and carried out the necessary surveys on wetwood because they are not aware that a problem exists: this stems largely from industry's reluctance to make a concerted effort to publicly recognize the problem.

The major reason the wetwood problem is not publicized is that most foresters and many wood processors have not related defects-- such as shake and frost cracks in trees and collapse and honeycomb in lumber--to wetwood. This is understandable because, until just recently, wood scientists have related the unexpected occurrence of honeycomb and ring failure when drying wetwood lumber to an inherent and perfectly normal variability in wood properties (75). Thus, lumber grading associations have been reluctant to recognize wetwood as a precursor to processing defects when even scientific experts are often unable to discern it in green lumber. In fact, there appears to be more recognition of wetwood and its problems by sawmill and dry kiln operators and furniture manufacturers than there is in the scientific community.

Logical species for the initial short-term studies are western hemlock, western true firs, eastern oaks, and the white pines, including sugar pine. These species are very susceptible to formation of wetwood and constitute a sizable volume of timber on commercial forest lands in the United States (192, 193). Comparisons of these species with other commercial species are shown in table 8. In spite of their susceptibility to wetwood, western hemlock, western true firs, eastern oaks, and white pines are tree species of high value and capable of producing valuable lumber items (table 9).

Table 8--Comparison of softwood and hardwood timber volumes from USDA Forest Service (192), and susceptibility of species to formation of wetwood

Species	Sawtimber volume		Growing stock volume		Wetwood class[1]	Study priority[2]
	Billion board feet	Percent	Billion cubic feet	Percent		
Western softwoods:						
Douglas-fir	520.640	27.3	96.861	22.4	1	
Western hemlock	251.012	13.2	47.540	11.0	3	1
True fir	218.772	11.5	45.326	10.5	2-3	2
Ponderosa and Jeffrey pine	189.897	10.0	38.292	8.9	1-2	
Spruce (Sitka, Engelmann, etc.)	132.225	6.9	26.296	6.1	1	
Lodgepole pine	65.273	3.4	25.530	5.9	1	
Sugar pine	23.520	1.2	4.344	1.0	2	4
Western white pine	20.872	1.1	3.993	.9	2	4
Western redcedar	40.897	2.2	8.106	1.9	2	
Western larch	31.256	1.6	6.753	1.6	2	
Redwood	23.627	1.2	4.428	1.0	2	
Other western species	31.362	1.7	6.886	1.6	1	
Total	1,549.353	81.3	314.355	72.8		
Eastern softwoods:						
Shortleaf and loblolly pine	196.502	10.3	53.571	12.4	1	
Longleaf and slash pine	44.248	2.3	13.855	3.2	1	
White and red pine	26.874	1.4	8.349	1.9	1-2	
Spruce and balsam fir	23.485	1.2	17.322	4.0	1-3	
Cypress	19.112	1.0	5.033	1.2	1	
Eastern hemlock	16.178	.9	5.781	1.3	3	
Other eastern species	29.534	1.6	13.611	3.2	1	
Total	355.933	18.7	117.522	27.2		
Total softwoods	1,905.286	100	431.877	100		
Eastern hardwoods:						
Red oak	106.217	20.6	39.309	18.1	2	3
White oak	78.689	15.3	32.099	14.8	1-2	3
Hickory	30.914	6.0	12.583	5.8	1-2	
Soft maple	23.871	4.6	15.070	6.9	2	
Hard maple	25.757	5.0	11.731	5.4	1	
Sweet gum	26.318	5.1	10.528	4.8	2	
Tupelo and black gum	25.506	5.0	9.817	4.5	2	
Yellow poplar	25.093	4.9	8.570	4.0	2	
Cottonwood and aspen	16.771	3.3	12.096	5.6	2-3	
Ash	15.957	3.1	7.736	3.6	2	
Beech	15.649	3.0	5.794	2.7	2	
Black cherry	6.904	1.3	3.488	1.6	1	
Basswood	8.502	1.6	3.434	1.6	1	
Yellow birch	7.324	1.4	3.249	1.5	2	
Other eastern species	46.313	9.0	22.178	10.2	1-2	
Total	459.785	89.2	197.682	91.1		
Western hardwoods:						
Red alder	24.842	4.8	7.638	3.5	1-2	
Cottonwood and aspen	12.077	2.3	5.043	2.3	2-3	
Oak	3.064	.6	1.606	.8	2	
Other western species	15.713	3.1	5.043	2.3	1-2	
Total	55.696	10.8	19.330	8.9		
Total hardwoods	515.481	100	217.012	100		

[1]/Class 1 species are least affected with wetwood; class 3 are most susceptible.

[2]/1 = highest; blanks indicate species with a priority lower than 4.

Table 9--Values of sawtimber stumpage and typical kiln-dried lumber prices for tree species with high priority for incorporation in a research program on wetwood

Species	Study priority[1]	Value of sawtimber stumpage[2]	Kiln-dried lumber		
			Grade	Thickness	1978 price[3]
		Dollars		Inches	Dollars per thousand board feet
Western softwoods:					
Western hemlock	1	46,245,409	Finish FG, C and Better selects	4/4	569
Mountain hemlock	1	344,395	No. 2 and Better, Hem-fir Dimension	8/4	244
Total		46,589,804			
White, grand, and					
miscellaneous firs	2	24,234,647			
Noble fir	2	5,046,826	Moulding and Better, rough dry	5/4, 6/4	526-562
Shasta red fir	2	2,097,222	No. 2 and Better Dimension	8/4	236
Subalpine fir	2	601,378			
Total		31,980,073			
Sugar and western white pine	4	40,004,600	C and Better Selects, S2S	4/4, 5/4, 6/4, 8/4	944-1,010
			Moulding, S2S	4/4, 5/4, 6/4, 8/4	463-761
Eastern hardwood:					
Oak	3	1,328,917	No. 1 Common and Better	4/4, 5/4	800-1, 100

[1]/1 = highest.
[2]/Value of sawtimber stumpage sold from National Forests in 1976 (143).
[3]/Averaged prices for softwood (Western Wood Products Association, Portland, Oreg.) and range of prices for oak (Hardwood Market Report, Volume 56, Memphis, Tenn.).

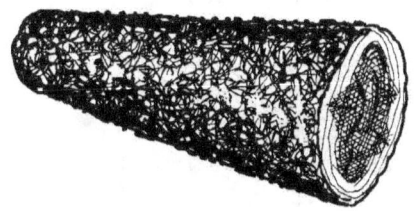

Material for the corollary studies on wetwood properties and processing should be selected from sample trees, logs, and wood products of the product and yield studies. Studies of wetwood properties should be designed to aid development of practical and accurate methods for detecting wetwood in timber and solid wood products in the green condition. Studies on processing methods should determine, both qualitatively and quantitatively, the wood losses, grade changes, and energy requirements for producing end products containing wetwood so that comparisons could be made with products containing normal wood. Results and feedback from the corollary studies would increase the efficiency and accuracy of timber quality evaluation and prediction of product potential for the commercial wood processor.

High priority should be given to three items of concern in the area of research to develop improved processing methods for timber and wood products containing wetwood:

1. Develop an accurate and fast method for identifying wetwood in green lumber and veneer to segregate the pieces into drying sorts.

2. Determine optimum methods for processing sorts of lumber and veneer containing pure and mixed amounts of wetwood and normal wood.

3. Compare mechanical strength properties of lumber containing wetwood with similar lumber of normal wood content for light frame construction.

There are good possibilities to design studies which could examine all three items.

Investigations to determine optimum methods of drying boards and veneers containing wetwood should probably be conducted in tandem with investigations to characterize and sort these products before they are dried. It would be useful to design these processing studies so that any development of drying defects could be related back to micro-organisms in the tree. The spread of bacteria from wetwood zones to sapwood during log storage needs to be investigated with respect to formation of chemical brown stain and changes in permeability of the sapwood. Definitive biochemical studies of contributing micro-organisms can also be considered a long-term research need.

A particular short-term research need must be the processing of boards and veneer containing mixtures of wetwood and normal wood. As was pointed out for white fir (147) and western hemlock (110), boards containing wood of a mixed drying sort present additional problems for drying and planing which have been ignored. Green boards and veneer containing mostly wetwood, if properly presorted, can be dried under special schedules. No data are presently available, however, for delineating characteristics of mixed wood sorts that would properly identify and permit presorting these products for drying. Techniques for drying mixed wood boards and veneer for optimum yields and energy consumption must be considered a short-term need of high priority.

The influence of wetwood on strength properties of lumber for light frame construction is not known. We can safely assume, though, that wetwood probably has a detrimental effect because of the frequent occurrence of deep surface checks, honeycomb, and shake in the dried lumber that reduces the resistance to shear. Shake is closely limited in stress-graded lumber for construction, particularly in members subject to bending (194). The data in table 7 suggest that wetwood may cause particleboard intended for structural purposes to have lower strength values than boards made entirely from normal wood. The importance of investigating the influence of wetwood in structural lumber and other building components cannot be overemphasized. According to Suddarth (184), the light-frame building is the predominant form of construction worldwide. Consequently, technological gains in light-frame construction can vitally affect large segments of the world's population.

Long-Term

The objectives of a long-term research program should be mainly concerned with attacking the causes of wetwood. Additional basic research aimed at reducing energy and wood losses during the processing of wetwood may also be needed. Specific plans for long-term research will depend somewhat on the initial information derived from short-term research, together with recommendations from workers in the fields of forest biology and timber management.

Knowledge of specific patterns and location of wetwood in standing trees gained largely from short-term research will provide the basis for planning long-term investigations of the causes of wetwood. The role of micro-organisms, especially anaerobic bacteria, need to be investigated. To be meaningful, though, research on microbiology must be coordinated with investigations of other possible causes. There is a need to investigate possible contributions by the following agents: insects, fungi, mistletoe, oxygen and water stress, fire, and mechanical wounds associated with timber management and recreational uses of the forest.

Identification and biochemical characterization of the bacterial populations associated with wetwood can help to solve current wood processing problems and provide a basis for future control of wetwood formation in trees. The biochemical contribution of pectolytic and phenyloxidizing bacteria to shake and frost cracks in trees and to surface checking, honeycomb, collapse, ring failure, and chemical brown stain in wood products needs to be more thoroughly defined and related to the initial formation of wetwood.

It is important to explore the possibility that bacteria in wetwood may enhance its subsequent decay by fungi if anaerobic conditions are lost, such as by frost cracks. Hartley et al. (79) noted that fungal wood decay can be accelerated by the presence of bacteria but did not consider the possibility that wetwood bacteria may provide a source of nitrogen and vitamins for fungal growth. Nitrogen-fixing bacteria have been considered in more recent papers[29] (3, 4, 22). Nitrogen fixation by bacteria has even been associated with sporophores of decay fungi on western hemlock trees (118). In addition, Bourchier (25) found thatbacteria from wetwood can produce vitamins which enhance the growth of wood-decaying fungi.

Studies of bacterial populations in wetwood must also be concerned with distinguishing inocuous plant and soil bacteria from pathogenic and fecal coliforms by serology, cell wall analysis, and deoxyribonucleic acid homology studies.[30] The presence of coliform bacteria in living wood tissues (13) can be a problem in some uses of wood products, such as redwood for water storage tanks (166).

---

[29]S. D. Spano, M. F. Jurgensen, M. J. Larsen, and A. E. Harvey. 1978. Nitrogen fixation in decaying Douglas fir residue. Paper presented at 70th Annual Meeting of the American Society of Agronomy, 11 p. Madison, Wis.

[30]J. G. Zeikus, University of Wisconsin, Department of Bacteriology, Madison, personal communication to J. C. Ward.

Once the causes of wetwood are understood, there are two methods for control: direct and indirect. Long-term research should evaluate the relative merits of each method. In the practice of forestry, direct control of tree diseases and insect injuries is generally not as practical or economical as indirect measures of control. Direct methods, such as chemical spraying, are more practical with agricultural crops than with long-term timber crops which have a lower value per unit of ground. Still, direct measures of control have been successfully used in forest management and must be considered for preventing wetwood. Tree species susceptible to wetwood during early stages of growth may require aerial spraying and soil fumigation. Special thinning methods and fertilization may also be needed in these young-growth stands, but wetwood is sometimes reported to be more prevalent in dominant, fast-growing trees than in suppressed or slow-growing trees (45, 59, 181).

The rotation age for sawtimber could be reduced for species that develop wetwood with advancing age. When guidelines are developed for this type of control, it is also necessary to determine the influence of site factors and genetic makeup of the tree species on its resistance or susceptibility to wetwood with aging. If wetwood develops in young trees before sawtimber size is reached, direct control measures could be aided by research on the utilization and processing of the timber. Pole-size or presawtimber stands might be converted into products of higher unit value than the present end uses for pulp and paper. The development of more efficient drying and machining methods could even allow young trees with wetwood to profitably grow to sawtimber size.

With current economic conditions, the goal of timber managers must be to grow sawtimber that will not develop wetwood before financial maturity. Indirect or preventive control measures necessary to achieve this goal can be obtained from forest biology and tree genetics research. This research must correlate the mechanisms of genetic resistance to wetwood formation with the influences of site and cultural practices. Benefits from a long-term research program could include:

1. Determination or development of trees that are genetically resistant to formation of wetwood.

2. Determination of growing conditions that will prevent or minimize formation of wetwood in susceptible tree species before the trees reach financial maturity for sawtimber.

3. Development of cultural and harvesting practices that will minimize the tendency for wetwood to form in residual trees or in regeneration.

Conclusion

From this overview, it is apparent that wetwood is responsible for substantial losses of wood, energy, and production expenditures in the forest products industry. These losses result from a raw material anomaly that in the past has either not been recognized or has been ignored. Though there is a definite need to assess more thoroughly these losses and to find ways of eliminating or minimizing them, there is also a need to examine the cause; i.e., wetwood. The fact that more and more problems in utilization and processing of timber are recognized as related directly or indirectly to wetwood clearly demonstrates the importance of this phenomenon.

An effective program of research is needed to determine the significance of wetwood in these problems. This program should plan and provide for obtaining both short- and long-term solutions to the overall problem. Because of increasing interest in wetwood, excellent opportunities now exist to initiate such a comprehensive research program.

There can be little doubt that the effects of wetwood on the forest products industry are real, and the impact is far reaching. With dwindling timber supplies, increasing costs of available energy, and a scarcity of low cost capital, the timber industry can no longer afford to ignore the problems associated with wetwood, for wetwood has an impact on all these issues.

It is time we examine, in depth, the detrimental effects of wetwood during all phases of timber production--from the stump to the finished product. Losses related to wetwood, which in the past were accepted as part of the package of doing business or thought to be the result of inadequate processing, can no longer be viewed in that context. A fuller understanding of the characteristics of wetwood, whether in the tree, log, or product, will provide timber managers and processors with the tools to handle the wetwood problem.

Literature Cited

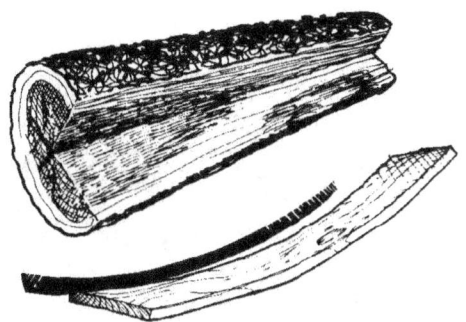

1. Abe, Z., and K. Minami.
    1976. [Ill smell from the wood of Gonystylus Bancanus (Miq.) Kurz. I. Source of ill smell. II. Smelling components.] J. Jap. Wood Res. Soc. (Mokuzai Gakkaishi) 22(2):119-122. [In Jap.]

2. Aho, P. E.
    1977. Decay of grand fir in the Blue Mountains of Oregon and Washington. USDA For. Serv. Res. Pap. PNW-229, 18 p. Pac. Northwest For. and Range Exp. Stn., Portland, Oreg.

3. Aho, P. E., and A. Hutchins.
    1977. Micro-organisms from the pith region of suppressed grand fir understory trees. USDA For. Serv. Res. Note PNW-299, 5 p. Pac. Northwest For. and Range Exp. Stn., Portland, Oreg.

4. Aho, P. E., R. J. Seidler, H. J. Evans, and P. N. Raju.
    1974. Distribution, enumeration, and identification of nitrogen-fixing bacteria associated with decay in living white fir trees. Phytopathology 64(11):1413-1420.

5. Anderson, A. B., E. L. Ellwood, E. Zavarin, and R. W. Erickson.
    1960. Influence of extractives on seasoning stain of redwood lumber. For. Prod. J. 10(4):212-218.

6. Anderson, A. B., and W. B. Fearing. 1961. Distribution of extractives in solvent seasoned redwood lumber. For. Prod. J. 11(5):240-242.

7. Arganbright, D. G., and W. A. Dost. 1972. The effect of brown stain on drying characteristics of sugar pine - preliminary observations. Proc. 23d Annu. Meet. West. Dry Kiln Clubs, p. 46-53. Oreg. State Univ., Corvallis.

8. Arganbright, D. G., and R. Smith. 1978. General drying characteristics of young-growth redwood. Tech. Rep. 35.01.156, 25 p. Univ. Calif. For. Prod. Lab., Richmond.

9. Arganbright, D. G., and W. W. Wilcox. 1969. Comparison of parameters for predicting permeability of white fir. Proc. Am. Wood-Preserv. Assoc. 65:57-62.

10. Arni, P. C., G. C. Cochrane, and J. D. Gray. 1965. The emission of corrosive vapours by wood. I. Survey of acid-release properties of certain freshly felled hardwoods and softwoods. J. Appl. Chem. 15(July):305-313.

11. Arni, P. C., G. C. Cochrane, and J. D. Gray. 1965. The emission of corrosive vapours by wood. II. The analysis of vapours emitted by certain freshly felled hardwoods and softwood by gas chromatography and spectrophotography. J. Appl. Chem. 15(Oct.):463-468.

12. Bacon, M., and C. E. Mead. 1971. Bacteria in the wood of living aspen, pine, and alder. Northwest Sci. 45(4):270-275.

13. Bagley, S. T., R. J. Seidler, H. W. Talbot, Jr., and J. E. Morrow. 1978. Isolation of Klebsielleae from within living wood. Appl. and Environ. Microbial. 36(1):178-185.

14. Baker, G. 1951. The effect of log storage time on development of brown stain in northern white pine (during kiln seasoning). Univ. Maine For. Dep. Tech. Note No. 8, 1 p. Orono.

15. Barton, G. M. 1968. Significance of western hemlock phenolic extractives in pulping and lumber. For. Prod. J. 18(5):76-80.

16. Barton, G. M. 1972. How to prevent dry kiln corrosion. Can. For. Ind. 92(4):27-29.

17. Barton, G. M. 1973. The significance of western hemlock phenolic extractives in groundwood pulping. Tappi 56(5):115-118.

18. Barton, G. M., and J. A. F. Gardner. 1966. Brown stain formation and the phenolic extractives of western hemlock (Tsuga heterophylla (Raf.) Sarg.) Dep. For. Publ. 1147, 20 p. Ottawa, Can.

19. Barton, G. M., and R. S Smith. 1971. Brown stain in kiln-dried Abies amabilis lumber. Bi-Mon. Res. Notes 27(3):21-23. Dep. Fish. and For., Ottawa, Can.

20. Bauch J., W. Höll, and R. Endeward. 1975. Some aspects of wetwood formation in fir. Holzforschung 29(6):198-205.

21. Bauch, J., W. Liese, and H. Berndt. 1970. Biological investigations for improvement of permeability of softwoods. Holzforschung 24(6):199-205.

22. Blanchette, R. A., and C. G. Shaw. 1978. Associations among bacteria, yeasts, and basidiomycetes during wood decay. Phytopathology 68(4):631-637.

23. Boisselle, H. J. 1969. Successful kiln drying of white pine for furniture stock. For. Prod. J. 19(3):17-20.

24. Boone, R. S., and J. C. Ward. 1977. Kiln drying lumber from American elm trees killed by Dutch elm disease. For. Prod. J. 27(5):48-50.

25. Bourchier, R. J. 1967. The role of bacterial wetwood in balsam fir decay. (Abstr.) Proc. 33d Session Can. Phytopathological Soc. 34:16 Ottawa.

26. Boutelje, J., and B. Ihlstedt. 1978. [The effects of wet storage of roundwood. 3. Effect on checking-tendency and sorption properties. 1 Medd. Sven. Traforskningsinstitutet A. 501, 21 p. [In Swed.].

27. Boyce, J. C. 1923. A study of decay in Douglas-fir in the Pacific Northwest. U.S. Dep. Agric. Bull= 1163, 20 p.

28. Bramhall, G., and W. G. Warren. 1977. Moisture content control in drying dimension lumber. For. Prod J. 27(7):26-28.

29. Bramhall, G., and R. W. Wellwood. 1976. Kiln drying of western Canadian lumber. Inf. Rep.VP-X-159, 112 p. Dep. Fish. and the Environment, Can. For. Serv. West. For. Prod. Lab., Vancouver, B.C.

30. Bryan, E. L. 1960. Collapse and its removal in Pacific madrone. For. Prod. J. 10(11):598-604.

31. Campbell, W. A., and R. W. Davidson. 1941. Redheart of paper birch. J. For. 39(1):63-65.

32. Carroll, M. N., and M. Dokken. 1970. Veneer drying problems in perspective. Can. Dep. Fish. and For. Inf. Rep. OP-X-32, 14 p. Ottawa, Can.

33. Carter, J. C. 1945. Wetwood of elms. Bull. Ill. Nat. Hist. Surv. 23(4):401-448.

34. Catterick, J. W., and G. B. Gillies. 1966. Production control of brown stain. For. Prod. J. 16(11):16.

35. Cech, M. Y. 1966. Brown stain in white pine. For. Prod. J. 16(11):23-27.

36. Cech, M. Y. 1973. The status of high temperature kiln-drying in eastern Canada. Can. For. Ind. 93(8):63-71.

37. Cech, M. Y. 1978. Low-tempetature kiln-drying: It works. Can. For. Ind. 98(2):20-22.

38. Chang, C. I. J. 1972. The cause of ring shake: A review of the literature. Q. J. Chin. For. 6(1):69-77.

39. Clarke, S. G., and E. E. Longhurst.
      1961. The corrosion of metals by
      acid vapours from wood. J. Appl.
      Chem. 11(11):435-443.

40. Clausen, V. H., and F. A. Kaufert.
      1952. Occurrence and probable cause
      of heartwood degradation in
      commercial species of _Poulus_. J.
      For. Prod. Res. Soc. 2(4):62-67.

41. Clausen, V. H., L. W. Rees, and F. H.
      Kaufert.
      1949. Development of collapse in
      aspen lumber. Proc. For. Prod. Res.
      Soc. 3:460-468.

42. Cole, E. J., Jr., and F. A. Streams.
      1970. Insects emerging from brown
      slime fluxes in southern New
      England. Can. Entomol.
      102(3):321-333.

43. Comstock, G. L.
      1975. Energy requirements for
      drying of wood products. _In_ Wood
      residues as an energy source. Proc.
      P-75-13, p. 8-12. For. Prod. Res.
      Soc., Madison, Wis.

44. Courtois, H.
      1966. [On the decomposition of the
      cell wall by bacteria in coniferous
      wood.] Holzforschung 20(5):148-154.
      [In German.)

45. Coutts, M. P., and J. Rishbeth.
      1977. The formation of wetwood in
      grand fir. Eur. J. For. Pathol.
      7(1):13-22.

46. Crandall, B. S.
      1943. Bacterial infection and decay
      of the inner wood of winter-injured
      young London plane trees.
      Phytopathology 33(10):963-964.

47. Day, W. R.
      1929. The watermark disease of the
      cricket-bat willow (Salix
      caerulea). Oxford For. Mem. 3, 30
      p. Oxford Univ. Press, Oxford,
      England.

48. Dokken, M., and R. Lefebvre.
      1973. Drying veneer peeled from
      seven New Brunswick balsam fir
      logs. Can. Dep. Environ. Inf. Rep.
      OP-X-60, 20 p. Ottawa.

49. Dowson, W. J.
      1937. _Bacterium salicis_ Day. The
      cause of the watermark disease of
      the cricket-bat willow. Ann. Appl.
      Biol. 24(3):528-544.

50. Dunleavy, J. A., and A. J. McQuire.
      1970. The effect of water storage
      on the cell-structure of Sitka
      spruce (_Picea sitchensis_) with
      reference to its permeability and
      preservation. J. Inst. Wood Sci.
      5(2):20-28.

51. Eades, H. W.
      1932. British Columbia softwoods:
      their decays and natural defects.
      For. Serv. Bull. 80, 126 p. Dep.
      Inter. Can. Ottawa.

52. Ecklund, B. A., W. A. Dost, and W. L.
Benjamin.
1962. The effect of pre-steaming on
stain control and drying rate of
redwood. Interim Rep. A, Index
3.2420, 14 p. Calif. Redwood Assoc.,
San Francisco.

53. Eis, S.
1972. Root grafts and their
silvicultural implications. Can. J.
For. Res. 2(2):111-120.

54. Ellwood, E. L., A. B. Anderson, E.
Zavarin, and R. W. Erickson.
1959. The effect of drying
conditions and certain pretreatments
on seasoning stain in California
redwood. Proc. 11th Annu. Meet.
West. Dry Kiln Clubs, p. 31-45.
Oreg. State Univ., Corvallis.

55. Ellwood, E. L., and B. A. Ecklund.
1959. Bacterial attack of pine logs
in pond storage. For. Prod. J.
9(9) :283-292.

56. Englerth, G. H., and J. R. Hansbrough.
1945. The significance of the
discolorations in aircraft lumber:
Noble fir and western hemlock. USDA
For. Pathol. Spec. Release 24, 10
p., illus. For. Prod. Lab.,
Madison, Wis.

57. Erdesi, J.
1969. [Frost crack and the
microbial metabolism of trees.]
Sumarstvo 22(7/8):3-19. [In Serb.]

58. Erickson, R., J. Haygreen, and R.
Hossfeld.
1966. Drying prefrozen redwood -
with limited data on other species.
For. Prod. J. 16(8):57-65.

59. Etheridge, D. E., and L. A. Morin.
1962. Wetwood formation in balsam
fir. Can. J. Bot. 40(10): 1335-1345.

60. Etheridge, D. E., and L. A. Morin.
1967. The microbiological condition
of wood of living balsam fir and
black spruce in Quebec. Can. J.
Bot. 45(7):1003-1010.

61. Evans, R. S., and H. N. Halvorson.
1962. Cause and control of brown
stain in western hemlock. For.
Prod. J. 12(8):367-373.

62. Fakirov, V.
1972. [Frost cracks in hybrid black
poplars on flooded and drained sites
along the Danube.] Gorsko
stop. 28(9):19-25. [In Bulg.]

63. Fogarty, W. M.
1973. Bacteria, enzymes, and wood
permeability. Process Biochem.
8(6) :30-34.

64. Ford-Robertson, F. C., Ed.
1971. Terminology of forest
science, technology practice and
products. 349 p. Soc. Am. For.,
Washington, D.C.

65. Foster, R. E., G. P. Thomas, and J. E.
Browne.
1953. A tree decadence
classification of mature coniferous
stands. For. Chron. 29(4):359-366.

66. Fujioka, M., and K. Takahashi.
1921. On the cause of the darkening
of the heartwood of Cryptomeria
japonica Don. J. For. 19(8):844-866.

67. Gertjejansen, R.
1969. Wet process hardboards from
aspen sapwood and discolored
heartwood. For. Prod. J.
19(9):103-107.

68. Gibbs, R. D.
    1935. Studies of wood. II. On the water content of certain Canadian trees and on changes in the water-gas system during seasoning and flotation. Can. J. Res. 12(6):727-760.

69. Greaves, H.
    1971. The bacterial factor in wood decay. Wood Sci. and Technol. 5(1):6-16.

70. Greaves, H.
    1973. Selected wood-inhabiting bacteria and their effect on strength properties and weights of Eucalyptus regnans F. Muell. Pinus radiata D. Don sapwoods. Holzforschung 27(1):20-26.

71. Grober, S.
    1942. The botanical, erosion control, and economic significance of white poplar in Maryland. Ph.D. thesis. Univ. Md., College Park, 46 p.

72. Grosu, R., M. Boiciuc, and R. Vladut.
    1973. [The influence of bacteria on the quality of wood and wood products. I. The influence of bacteria on wood and wood stored in water.] Ind. Lemnului 24(4):143-147. [In Romanian.]

73. Gurensey, F. W.
    1951. Collapse in western red cedar. B. C. Lumberman 35(4):44-45, 62.

74. Hadzi-Georgiev, K., and N. Popnikola.
    1971. [Stem cracks in poplar in Macedonia.] Topola 15(83/85): 49-54. [In Serb.]

75. Hann, R. A., and J. C. Ward.
    1972. A look at and a result of reorganizing drying research at Forest Products Laboratory. Proc. 23d Annu. Meet. West. Dry Kiln Clubs, p. 20-25. Oreg. State Univ., Corvallis.

76. Harmsen, L., and T. V. Nissen.
    1965. [Bacterial attacks on wood.) Holz Als Roh-Und Werkst. 23(1):389-393. [In Ger.]

77. Hartig, R.
    1839. [Text-book of the diseases of trees.] Translated by William Somerville, 1894. Rev., 331 p. Macmillan and Co., New York.

78. Hartley, C., and R. W. Davidson.
    1950. Wetwood in living trees. (Abstr.) Phytopathology 40(9):871.

79. Hartley, C., R. W. Davidson, and B. S. Crandall.
    1961. Wetwood, bacteria, and increased pH in trees. USDA For. Prod. Lab. Rep. 2215, 35 p. Madison, Wis.

80. Haygreen, J. G., and S.-S. Wang.
    1966. Some mechanical properties of aspen wetwood. For. Prod. J. 16(9):118-119.

81. Heiskanen, V.
    1955. Water-core as a factor diminishing the value of pine logs. For. Abstr. 16(4):582, No. 4732.

82. Herpka, I., N. Zivanov, and J. Markovic.
    1969. [Frost crack in poplar stems and the results of research to avoid it.] Topola 13(75/76):5-34. [In Serb.]

83. Hillis, W. E., Y. Yazaki, and J. Bauch.
    1976. The significance of anomalous extractives in heart-shakes in Dacrydium species. Wood Sci. and Technol. 10(2):79-95.

84. Hossfeld, R. L., J. C. Oberg, and D. W. French.
    1957. The appearance and decay resistance of discolored aspen. For. Prod. J. 7(10):378-382.

85. Hubert, E. E.
    1926. The brown stains of lumber. Timberman 27(7):44-45, 48, 50.

86. Huffman, D. R.
    1972. Kiln-drying aspen studs. For. Prod. J. 22(10):21-23.

87. Hunt, D. L.
    1963. Seasoning and surfacing degrade in kiln-drying western hemlock in western Washington. Proc. 15th Annu. Meet. West. Dry Kiln Clubs, p. 32-37. Oreg. State Univ., Corvallis.

88. Ishida, S.
    1963. [On the development of frost cracks on "Todomatsu," Abies sachalinensis, trunks especially in relation to their wetwood.] Res. Bull. Coll. Exp. Stn. Hokkaido Univ. 22(2):273-373 + 25 plates. [In Jap.]

89. Jenkins, J. H.
    1934. Kiln-drying British Columbia lumber. For. Serv. Bull. 86, 78 p. For. Prod. Lab, Can. Dep. Inter., Ottawa.

90. Johnson, B. R.
    1979. Permeability changes induced in three western conifers by selective bacterial inoculation. Wood and Fiber 11(1):10-21.

91. Johnson, N. E., and K. R. Shea.
    1963. White fir defects associated with attacks by the fir engraver. For. Res. Note 54, 8 p. Weyerhaeuser Co., Centralia, Wash.

92. Kaburagi, J.
    1973. [On the moisture content of green Todo-Fir trunks (Abies sachalinensis), especially in relation to their wetwood.] Bull. Exp. For., Tokyo Univ. Agric. and Technol. 10, p. 96-107. [In Jap.]

93. Kamp, B. J. van der, A. A. Gokhale, and R. S. Smith.
    1979. Decay resistance owing to near-anaerobic conditions in black cottonwood wetwood. Can. J. For. Res. 9(1):39-44.

94. Karnop, G.
    1972. [Cellulose decomposition and destructive patterns on some wood components of water-stored softwoods caused by Clostridium omelianski.] Mater. und Org. 7(3):189-203. [In Ger.]

95. Karnop, G.
    1972. [Morphology, physiology, and destructive patterns of non-celluloytic bacteria of water-stored softwoods.] Mater. und Org. 7(2):119-132. [In Ger.]

96. Kaufert, F. H.
    1975. Minnesota's native wild elms. Univ. Minn. Agric. Exp. Stn. Misc. Rep. 131, 12 p. St. Paul, Minn.

97. Keays, J. L.
        1977. Forest harvesting of the
        future. AIChE Symp. Ser.
        72(157):4-12. Am. Inst. Chem. Eng.

98. Kemp, A. K.
        1959. Factors associated with the
        development of collapse in aspen
        during kiln drying. For. Prod. J.
        9(3):124-130.

99. Knauss, A. C., and E. H. Clarke.
        1961. Seasoning and surfacing
        degrade in kiln-drying ponderosa
        pine in eastern Washington. USDA
        For. Serv. Pac. Northwest For. and
        Range Exp. Stn. Res. Note 205,
        12 p. Portland, Oreg.

100. Knauss, A. C., and E. H. Clarke.
        1961. Seasoning and surfacing
        degrade in kiln-drying western
        hemlock in western Oregon. USDA
        For. Serv. Pac. Northwest For. and
        Range Exp. Stn. Res. Note 207,
        11 p. Portland, Oreg.

101. Knight, E.
        1955. Identification of white fir
        drying sorts. Western Pine Assoc.
        Res. Note 4.213, 5 p.
        Portland, Oreg.

102. Knight, E.
        1959. The elimination of sugar
        pine brown stain. Proc. 11th Annu.
        Meet. West. Dry Kiln Clubs,
        p. 25-30. Oreg. State Univ.,
        Corvallis.

103. Knight, E.
        1970. Kiln drying western
        softwoods. Bull. 7004, 77 p.
        Moore Dry Kiln Co. Oreg.,
        Portland, Oreg.

104. Knuth, D. T.
        1964. Bacteria associated with
        wood products and their effects on
        certain chemical and physical
        properties of wood. Ph.D.
        thesis. Univ. Wis., Madison,
        186 p.

105. Knuth, D. T., and E. McCoy.
        1962. Bacterial deterioration of
        pine logs in pond storage. For.
        Prod. J. 12(9):437-442.

106. Knutson, D. M.
        1968. Wetwood in trembling aspen
        (Populus tremuloides Michx.). Ph.D.
        thesis. Univ. Minn., St. Paul.
        152 p.

107. Knutson, D. M.
        1973. The bacteria in sapwood,
        wetwood and heartwood of trembling
        aspen. Can. J. Bot. 51(2):498-500.

108. Kollmann, F. F. P., and W. A. Coté, J
        1968. Principles of wood science
        and technology. I. Solid wood.
        592 p. Springer-Verlag, New York.

109. Kozlik, C. J.
        1971. Electrical moisture meter
        readings on western hemlock
        dimension lumber. For. Prod. J.
        21(6):34-35.

110. Kozlik, C. J., and L. W. Hamlin.
        1972. Reducing variability in
        final moisture content of kiln
        dried western hemlock lumber. For
        Prod. J. 22(7):24-31.

111. Kozlik, C. J., R. L. Krahmer, and R.
        T. Lin.
        1972. Drying and other related
        properties of western hemlock
        sinker heartwood. Wood and Fiber
        4(2):99-111.

112. Krahmer, R. L., R. W. Hemingway, and W. E. Hillis.
1970. The cellular distribution of lignans in Tsuga heterophylla wood. Wood Sci. and Technol. 4(2):122-139.

113. Kubler, H.
1959. [Studies on growth stresses in trees.] Part I. The origin of growth stresses and stresses in transverse direction.] Holz Roh Werkst. 17(1):1-9. [In Ger.]

114. Kubler, H.
1959. [Studies on growth stresses in trees. Part II. Longitudinal stresses.] Holz Roh Werkst. 17(2):44-54. [In Ger.]

115. Kubler, H.
1959. [Studies on growth stresses in trees. Part III. Effect of heat treatment on the dimensions of green wood.] Holz Roh Werkst. 17(3):77-86. [In Ger.]

116. Kutsche, N. P., and R. L. Ethington.
1962. Shelling failures. For. Prod. J. 12(11):538.

117. Lagerberg, T.
1935. Barrtgrädens Vattved. [Wetwood in conifers.] Sven. Skogsvardsforeningens Tidskr. 33(3):177-264.

118. Larsen, M. J., M. F. Jurgensen, A. E. Harvey, and J. C. Ward.
1978. Dinitrogen fixation associated with sporophores of Fomitopsis pinicola, Fomes fomentarius, and Echinodontium tinctorum. Mycologia 70(6):1217-1222.

119. Liese, W., and G. Karnop.
1968. [On the attack of coniferous wood by bacteria.] Holz Roh Werkst. 26(6):202-208. [In Ger.]

120. Lin, R. T., and C. J. Kozlik.
1971. Permeability and drying behavior of western hemlock. Proc. 22d Annu. Meet. West. Dry Kiln Clubs, p. 44-50. Oreg. State Univ., Corvallis.

121. Lin, R. T., E. P. Lancaster, and R. L. Krahmer.
1973. Longitudinal water permeability of western hemlock. I. Steady-state permeability. Wood and Fiber 4(4):278-289.

122. Linzon, S. N.
1958. Water content variation in the heartwood of white pine and its relation to incipient decay. For. Chron. 34(1):48-49.

123. Linzon, S. N.
1962. Artificial inoculation of wet and dry heartwood of living eastern white pine trees. For. Sci. 8(2):163-167.

124. Lockard, C. R., J. A. Putnam, and R. D. Carpenter.
1963. Grade defects in hardwood timber and logs. U.S. Dep. Agric. Agcic. Handb. 244, 39 p.

125. Lutz, J. F.
1971. Wood and log characteristics affecting veneer production. USDA For. Serv. Res. Pap. FPL-150, 31 p. For. Prod. Lab., Madison, Wis.

126. Lutz, J. F.
1972. Veneer species that grow in the United States. USDA For. Serv. Res. Pap. FPL-167, 129 p. For. Prod. Lab., Madison, Wis.

127. McGinnes, E. A., Jr.
1965. Extent of shake in Missouri oaks. For. Prod. J. 15(5):190.

128. McGinnes, E. A., Jr.
1968. Extent of shake in black walnut. For. Prod. J. 18(5):80-82.

129. McGinnes, E. A., Jr., C. I. J. Chang, and K. Y. T. Wu.
1971. Ring shake in some hardwood species. Individual tree approach. J. Polym. Sci., Part C., No. 36, p. 153-176. Interscience Publ., New York.

130. McGinnes, E. A., Jr., J. E. Phelps, and J. C. Ward.
1974. Ultrastructure observations of tangential shake formations in hardwoods. Wood Sci. 6(3):206-211.

131. McIntosh, J. A., and T. Szabo.
1972. Watch for these signs of white spruce heartshake. Can. For. Ind. 92(10):68-69, 71.

132. MacKay, J. F. G.
1974. High-temperature kiln-drying of northern aspen 2- by 4-inch light-framing lumber. For. Prod. J. 24(10):32-35.

133. MacKay, J. F. G.
1975. Properties of northern aspen discolored wood related to drying problems. Wood and Fiber 6(4): 319-326.

134. MacKay, J. F. G.
1976. Delayed shrinkage after surfacing of high-temperature kiln-dried northern aspen dimension lumber. For. Prod. J. 26(2):33-36.

135. MacLean, H., and J. A. F. Gardner.
1951. Deterioration of wooden dry kilns used for drying western hemlock lumber. Lumberman 78(12):88-90.

136. MacLean, H., and J. A. F. Gardner.
1956. Distribution of fungicidal extractives (Thujaplicin and water-soluble phenols) in western redcedar heartwood. For. Prod. J. 6(12):510-516.

137. McMillen, J. M.
1953. Kiln drying water and swamp tupelo. J. For. Prod. Res. Soc. 3(5):189-196.

138. Manson, B. C.
1949. The drying of California redwood. Calif. Redwood Assoc. Res. Rep. 1, 8 p. San Francisco.

139. Meyer, R. W., and G. M. Barton.
1971. A relationship between collapse and extractives in western redcedar For. Prod. J. 21(4):58-60

140. Meyer, R. W., and L. Leney.
1968. Shake in coniferous wood: An anatomical study. For. Prod. J. 18(2):51-56.

141. Millett, M. M.
1952. Chemical brown stain in sugar pine. J. For. Prod. Res. Soc. 2(5):232-236.

142. Panshin, A. J., and C. de Zeeuw.
1970. Text book of wood technology. Vol. I. 3d ed. 705 p. McGraw-Hill Book Co., New York.

143. Phelps, R. B.
1977. The demand and price situation for forest products, 1976-77. U.S. Dep. Agric. Misc. Publ. 1357, 95 p.

144. Poluboyarinov, O. I.
1963. [The nature and some properties of aspen.] Nauchn. Tr. Leningr. Lesotekh. Akad. 102:37-44. [In Russ.]

145. Pong, W. Y.
1971. Changes in grade and volume of central California white fir lumber during drying and surfacing. USDA For. Serv. Res. Note PNW-156, 20 p. Pac. Northwest For. and Range Exp. Stn., Portland, Oreg.

146. Pong, W. Y., and G. H. Jackson.
1971. Diagraming surface characteristics of true fir logs. Supplement to log diagraming guide for western softwoods. USDA For. Serv. Pac. Northwest For. and Range Exp. Stn., 7 p., Portland, Oreg.

147. Pong, W. Y., and W. W. Wilcox.
1974. Spatial distribution of lumber degrade in white fir trees. USDA For. Serv. Res. Pap. PNW-184, 18 p. Pac. Northwest For. and Range Exp. Stn., Portland, Oreg.

148. Pratt, M. B.
1915. The deterioration of lumber. Calif. Agric. Exp. Stn. Bull. 252, p. 301-320, Berkeley.

149. Random Lengths.
1976. yearbook. Vol. 12. 118 p. David S. Evans, ed. Random Lengths Publ., Inc., Eugene, Oreg.

150. Random Lengths.
1977. Yearbook. Vol. 13. 186 p. David S. Evans, ed. Random Lengths Publ., Inc., Eugene, Oreg.

151. Rasmussen, E. F.
1961. Dry kiln operators manual. U.S. Dep. Agric. Agric. Handb. 188, 197 p.

152. Resch, H., E. T. Choong, and H. H. Smith.
1968. Sorting incense cedar types for drying. For. Prod. J. 18(4):40-44.

153. Resch, H., and B. A. Ecklund.
1964. Accelerating the drying of redwood lumber. Calif. Agric. Exp. Stn. Bull. 803, 27 p. Univ. Calif. Div. Agric. Sci., Berkeley.

154. Resch, H., B. A. Ecklund, and D. R. Prestemon.
1963. Tanoak drying program and shrinkage characteristics. Calif. Agric. (Oct.), p. 12-14. Univ. Calif. Div. Agric. Sci., Berkeley.

155. Rietz, R. C., and R. H. Page.
1971. Air drying of lumber: A guide to industry practices. U.S. Dep. Agric. Agric. Handb. 402, 110 p.

156. Rosen, H. N.
1978. High temperature predrying of wood. Proc. 29th Annu. Meet. West. Dry Kiln Clubs, p. 1-10. Oreg. State Univ., Corvallis.

157. Rossell, S. E., E. G. M. Abbot, and J. F. Levy.
1973. A review of the literature relating to the presence, action, and the interaction of bacteria in wood. J. Inst. Wood Sci. 6(2):28-35.

158. Sachs, I. B., J. C. Ward, and R. E. Kinney.
1974. Scanning electron microscopy of bacterial wetwood, sapwood, and normal heartwood in poplar trees. Proc. 7th Annu. Scanning Electron Microsc. Symp., Part 2, p. 453-460. IIT Res. Inst., Chicago.

159. Salamon, M.
     1961. Kiln drying of British
     Columbia softwoods at high
     temperatures. Proc. 13th Annu.
     Meet. West. Coast Dry Kiln Clubs,
     p. 29-35. Oreg. State Univ.,
     Corvallis.

160. Salamon, M.
     1973. Drying of lodgepole pine and
     spruce studs cut from flooded
     timber: A progress report. Proc.
     24th Annu. Meet. West. Dry Kiln
     Clubs, p. 51-56. Oreg. State
     Univ., Corvallis.

161. Salamon, M., and K. Urquhart.
     1972. 55-hour kiln schedule for
     1½-in. dry spruce. Can. For.
     Ind. 92(5):67, 69.

162. Schafer, R. F.
     1974. How to kiln-dry mixed stock
     for uniform moisture content. Can.
     For. Ind. 94(3):67-68.

163. Schirp, M.
     1968. Frostrisse an baumstämmen.
     Forstarchiv 39(7):149-154.

164. Schrenk, H. von.
     1905. Glassy fir. Rep. Mo. Bot,
     Gard. Annu. 16, p. 117-120. St.
     Louis, Mo.

165. Schroeder, H. A., and C. J. Kozlik.
     1972. The characterization of
     wetwood in western hemlock. Wood
     Sci. and Technol. 6(2):85-94.

166. Seidler, R. J., J. E. Morrow, and S.
     T. Bagley.
     1977. Klebsielleae in drinking
     water emanating from redwood
     tanks. Appl. and Environ.
     Microbiol. 33(4):893-900.

167. Seliskar, C. E.
     1950. Some investigations on the
     wetwood diseases of American elm
     and Lombardy poplar. Ph.D.
     thesis. Cornell Univ., Ithaca, N.Y.
     137 p.

168. Seliskar, C. E.
     1952. Wetwood organism in aspen,
     poplar is isolated. Colo. Farm and
     Home Res. 2(6):6-11, 19-20. Colo.
     State Univ. Fort Collins.

169. Sherrard, E. C., and E. F. Kurth.
     1933. The distribution of
     extractives in redwood: Its
     relation to durability. Ind. Eng.
     Chem. 25(3):300-302.

170. Shields, J. K., R. L. Desai, and M.
     R. Clarke.
     1973. Control of brown stain in
     kiln-dried eastern white pine.
     For. Prod. J. 23(10):28-30.

171. Shigo, A. L.
     1967. Successions of organisms in
     discoloration and decay of wood.
     Int. Rev. For. Res. 2:237-299.

172. Shigo, A. L.
     1972. Ring and ray shakes
     associated with wounds in trees.
     Holzforschung 26(2):60-62.

173. Shigo, A. L., and W. E. Hillis. 1973. Heartwood, discolored wood, and microorganisms in living trees. Annu. Rev. Phytopathol. 11:197-222.

174. Shigo, A. L., J. Stankewich, and B. J. Cosenga. 1971. *Clostridium* sp. associated with discolored tissue in living oaks. Phytopathology 61(1) :122-123.

175. Simpson, W. T. 1975. Effect of steaming on the drying rate of several species of wood. Wood Sci. 7(3):247-255.

176. Smith, H. H. 1956. Improved utilization of western hardwoods by modern drying. For. Prod. J. 6(3):121-124.

177. Smith, H. H., and J. R. Dittman. 1960. Drying rate of white fir by segregations. USDA For. Serv. Pac. Southwest For. and Range Exp. Stn. Res. Note 168, 10 p. Berkeley, Calif.

178. Smith, H. H., and J. R. Dittman. 1960. Moisture content in kiln-dried lumber. For. Prod. J. 10(7):353-357.

179. Smith, H. H., and J. R. Dittman. 1960. The segregation of white fir for kiln drying. USDA For. Serv. Pac. Southwest For. & Range Exp. Stn. Res. Note 167, 6 p. Berkeley, Calif.

180. Smith, R. S. 1975. Economic aspects of bacteria in wood. *In* Proceedings of sessions on wood product's pathology and 2nd International Congress of Plant Pathology, Sept. 10-12, 1973, Minneapolis, Minn. Springer-Verlag, New York.

181. Stojanov, V., and E. Enthev. 1968. [Properties of silver fir wetwood.] Gorskostop Nauka Sofija 5(5):61-73. [In Bulgarian.]

182. Stutz, R. E. 1959. Control of brown stain in sugar pine with sodium azide. For. Prod. J. 9(12):459-463.

183. Stutz, R. E., and A. W. Stout. 1957. The nature of brown stains in timber from the western pines. Proc. Plant Physiol. Meet. Plant Physiol. V 32, Suppl. 13, (Aug. 25). Am. Soc. Plant Physiol., Bethesda, Md.

184. Suddarth, S. K. 1973. Research needs in light-frame construction. Res. Bull. 903, 57 p. Wood Res. Lab., Purdue Univ., Lafayette, Ind.

185. Svidenko, A. I. 1970. [Damage to *Juglans regia* by frost cracks in the Bukovina region (Ukraine).] Lesoved., Mosk. 1970(1):89-90. [In Russ.]

186. Takizawa, T., N. Kawaguchi, M. Takahashi, and H. Yamamoto. 1976. [The observation of the wetwood of Todomatsu (*Abies* sachalinesis Mast.)] J. Hokkaido For. Prod. Res. Inst. 294(July):6-11. [In Jap.]

187. Terashima, N. 1975. Odor problems. Mokuzai Kog. 30(11):32-34. Wood Tech. Assoc. Jap., Tokyo 105. [In Jap.]

188. Thomas, D. P., and H. D. Erickson.
1963. Collapse and honeycomb in
western red cedar in relation to
green-wood liquid permeability.
Proc. 15th Annu. Meet. West. Coast
Dry Kiln Clubs, p. 7-14. Oreg.
State Univ., Corvallis.

189. Thunell, B.
1947. On the technical properties
of wetwood. For. Abstr.
8(3):400-401.

190. Tiedemann, G., J. Bauch, and E. Bock.
1977. Occurrence and significance
of bacteria in living trees of
Populus nigra L. Eur. J. For.
Pathol. 7(6):364-374.

191. Toole, E. R.
1968. Wetwood in cottonwood.
Plant Dis. Rep. 52(10):822-823.

192. U.S. Department of Agriculture,
Forest Service.
1973. The outlook for timber in
the United States. USDA For. Serv.
For. Resour. Rep. FRR-20, 367 p.,
illus. U.S. Gov. Print. Off.,
Washington, D.C.

193. U.S. Department of Agriculture,
Forest Service.
1977. The Nation's renewable
resources: An assessment, 1975.
USDA For. Serv. For. Resour. Rep.
FRR-21, 243 p. U.S. Gov. Print.
Off., Washington, D.C.

194. U.S. Forest Products Laboratory.
1974. Wood handbook: Wood as an
engineering material. U.S. Dep.
Agric. Agric. Handb. 72.

195. Unligil, H. H.
1969. Penetrability of white
spruce wood after water storage. J.
Inst. Wood Sci. 5(6):30-35.

196. Wagener, W. W.
1970. Frost cracks: A common
defect in white fir in California.
USDA For. Serv. Res. Note PSW-209,
5 p. Pac. Southwest For. and Range
Exp. Stn., Berkeley, Calif.

197. Wallin, W. B.
1954. Wetwood in balsam poplar.
Minn. For. Notes 28, Agric. Exp.
Stn. Sci. J. Ser. Pap. 3118, 2 p.
Univ. Minn., St. Paul.

198. Wangaard, F. F.
1978. Anatomy and behavior of
early wood in relation to water
stress in living trees. Abstr.,
32d Annu. Meet. For. Prod. Res.
Soc., p. 33. Madison, Wis.

199. Ward, J. C.
1972. Anaerobic bacteria
associated with honeycomb and ring
failure in red and black oak
lumber. (Abstr.) Phytopathology
62(7):796.

200. Ward, J. C.
1976. Kiln drying characteristics of studs from Rocky Mountain aspen and Wisconsin aspen. In Utilization and marketing as tools for aspen management in the Rocky Mountains. USDA For. Serv. Gen. Tech. Rep. RM-29, p. 73-74. Rocky Mt. For. and Range Exp. Stn., Fort Collins, Colo.

201. Ward, J. C.
1978. Honeycomb: Dry kiln or bacteria. Abstr., 32d Annu. Meet. For. Prod. Res. Soc., p. 30. Madison, Wis.

202. Ward, J. C., R. A. Hann, R. C. Baltes, and E. H. Bulgrin.
1972. Honeycomb and ring failure in bacterially infected red oak lumber after kiln drying. USDA For. Serv. Res. Pap. FPL-165, 36 p., illus. For. Prod. Lab., Madison, Wis.

203. Ward, J. C., and C. J. Kozlik.
1975. Kiln drying sinker heartwood from young-growth western hemlock: Preliminary evaluation. Proc. 26th Annu. Meet. West. Dry Kiln Clubs, p. 44-63. Oreg. State Univ., Corvallis.

204. Ward, J. C., J. E. Kuntz, and E. McCoy.
1969. Bacteria associated with "shake" in broadleaf trees. (Abstr.) Phytopathology 59(8):1056.

205. Ward, J. C., and D. Shedd
1979. California black oak drying problems and the bacterial factor. USDA For. Serv. Res. Pap. FPL-344, 14 p. For. Prod. Lab., Madison, Wis.

206. Ward, J. C., and E. M. Wengert.
1974. Bacterial wetwood: How it affected wood drying quality in aspen. Proc. 3d North Am. For. Biol. Workshop, p. 382. Colo. State Univ., Fort Collins.

207. Ward, J. C., M. F. Wesolowski, and A. L. Shigo.
1978. The use of electrical resistance for detecting hard-to-dry lumber containing bacterial wetwood. Abstr., 32d Annu. Meet. For. Prod. Res. Soc., p. 36. Madison, Wis.

208. Ward, O. P., and W. M. Fogarty.
1973. Bacterial growth and enzyme production in Sitka spruce (Picea sitchensis) sapwood during water storage. J. Inst. Wood Sci. 6(2):.8-12.

209. Warren-Wren, S. C.
1965. The significance of the caerulean or cricket-bat willow (Salix alba L. cultivar "calva"). Q. J. For. 59(3):193-205.

210. Western Wood Products Association.
1977. Grading rules for western lumber. 4th ed. 202 p., illus. Portland, Oreg.

211. Wickman, B. E., and R. F. Scharpf.
1972. Decay in white fir top-killed by Douglas-fir tussock moth. USDA For. Serv. Res. Pap. PNW-133, 9 p. Pac. Northwest For. and Range Exp. Stn., Portland, Oreg.

212. Wilcox, W. W.
     1968. Some physical and mechanical
     properties of wetwood in white
     fir. For. Prod. J. 18(120):27-31.

213. Wilcox, W. W.
     1970. Anatomical changes in wood
     cell walls attacked by fungi and
     bacteria. Bot. Rev. 36(1):1-28.

214. Wilcox, W. W., and N. D. Oldham.
     1972. Bacterium associated with
     wetwood in white fir.
     Phytopathology 62(3):384-385.

215. Wilcox, W. W., and W. Y. Pong.
     1971. The effects of height,
     radial position, and wetwood on
     white fir wood properties. Wood
     and Fiber 3(1):47-55.

216. Wilcox, W. W., W. Y. Pong, and J. R.
     Parmeter.
     1973. Effects of mistletoe and
     other defects on lumber quality in
     white fir. Wood and Fiber
     4(4):272-277.

217. Wilcox, W. W., and C. G. R. Schlink.
     1971. Absorptivity and pit
     structure as related to wetwood in
     white fir. Wood and Fiber
     2(4):373-379.

218. Williston, E. M.
     1971. Drying west coast dimension
     to meet the new lumber standards.
     For. Prod. J. 21(3):44-48.

219. Wong, W. C., and T. F. Preece.
     1978. *Erwinia salicis* in cricket
     bat willows: Histology and
     histochemistry of infected woods.
     Physiol. Plant Pathol.
     12(3):321-332.

220. Yazawa, K., S. Ishida, and H.
     Miyajima.
     1965. On the wet-heartwood of some
     broad-leaved trees grown in Japan.
     I. J. Jap. Wood Res. Soc.
     (Mokuzai Gakkaishi) 11(3):71-76.

221. Zeikus, J. G., and D. L. Henning.
     1975. *Methanobacterium*
     arbophilicum sp. nov. An obligate
     anaerobe isolated from wetwood of
     living trees. Antonie van
     Leeuwenhoek J. Microbial. and
     Serol. 41(4):543-552.

222. Zeikus, J. G., and J. C. Ward.
     1974. Methane formation in living
     trees: A microbial origin.
     Science 184(4142):1181-1183.

223. Zinkel, D. F., J. C. Ward, and B. F.
     Kukachka.
     1969. Odor problems from some
     plywoods. For. Prod. J. 19(12):60.